# THE GLOBALIZATION MYTH

## WHY THE PROTESTORS HAVE GOT IT WRONG

ALAN SHIPMAN

Published in the UK in 2002 by
Icon Books Ltd., Grange Road,
Duxford, Cambridge CB2 4QF
E-mail: info@iconbooks.co.uk
www.iconbooks.co.uk

Sold in the UK, Europe, South Africa
and Asia by Faber and Faber Ltd.,
3 Queen Square, London WC1N 3AU
or their agents

Distributed in the UK, Europe,
South Africa and Asia by
TBS Ltd., Frating Distribution
Centre, Colchester Road, Frating
Green, Colchester CO7 7DW

Published in Australia in 2002
by Allen & Unwin Pty. Ltd.,
PO Box 8500, 83 Alexander Street,
Crows Nest, NSW 2065

Reprinted 2003

Published in the United States
in 2004 by Totem Books
Inquiries to Icon Books Ltd.,
Grange Road, Duxford,
Cambridge CB2 4QF, UK

In the United States, distributed
to the trade by National Book
Network Inc., 4720 Boston Way,
Lanham, Maryland 20706

Distributed in Canada by
Penguin Books Canada,
10 Alcorn Avenue, Suite 300,
Toronto, Ontario M4V 3B2

ISBN 1 84046 359 7



Typeset by Wayzgoose

Printed and bound in the UK by
Cox & Wyman Ltd., Reading

# Contents

## Acknowledgements

June Edmunds and Josephine Donnelly Edmunds endured the anti-social fallout from this work's high-speed production, and contributed more insights and ideas than either might wish to recognize. Peter Nolan and Geoffrey Harcourt provided an immediate gateway to the project, and lasting input to its underlying approach. Duncan Heath, Andrew Furlow, Jeremy Cox and Peter Pugh at Icon Books supported the original idea, showed patience with the changes of plan, and worked wonders to keep the whole thing in shape and on schedule. Fiona McMorrough bravely undertook to find markets for a book that is often sceptical about them. Conversations with Sonia Bhalotra, Ayalvadi Ganesh, Jonathan Koehler, Philip Lucien, Alex Parr, Cath Porter, Margaret and Marten Shipman were also useful in refining various arguments.

More generalized thanks are due to members of the Critical Realist Workshop in Cambridge and the Prospect Reading Group in London, for keeping alive the belief that arguments rooted in 'social science' are still worth producing and consuming. Any crisis in that faith provoked by what follows is unintentional, if at times inevitable – rather like the collateral damage from globalization itself.

**Alan Shipman** is a freelance commentator who writes and broadcasts extensively on international finance and politics, emerging markets, global business strategy and the foundations of the market economy. After studying economics at two ancient universities, and finding it not much use for survival in the real economy, he researched and consulted for a number of distinguished organizations before starting to come up with solutions of his own. Global pressure to raise productivity and please editors have supposedly eliminated spare-time activities; but these include nocturnal jogging, escaping domestic responsibilities, ghost-writing for unknown heads of state, and improbable bicycle journeys.

He is also the author of *The Market Revolution and its Limits* (1999) and *Transcending Transaction* (2001), both published by Routledge.

## INTRODUCTION

# Planetary Spin

Life on earth was global from the outset, as one fragile planet huddled for comfort against a cold and empty space. Localization came later, after manners started to fragment over space, and memories over time. Many efforts have since been made to turn back the dispersing tide and restore our cross-border connections. This book explores the motives, mechanisms and myths that surround our latest plunge towards a planet without frontiers.

Early attempts at renewed global unity were dictatorial and personal. Solidarity meant subjection to a common set of rules, and the rule-makers were known by name. Alexander the Great's triumphant lament of having no new worlds to conquer was repeated by Holy Roman Emperors Charlemagne and Charles V; with unamused rejoicing by Queen Victoria, and with premature celebration by general secretary Joseph Stalin. The most recent approach to global personal dictatorship, that imposed on the world's computer screens by Bill Gates, changes the pattern but not the underlying program.

Initial border-busting moves were also founded on exclusion. Piecing the whole world back together soon became too daunting as language, culture, politics and economic interests went their separate ethnographic ways. So the bridge-builders concentrated on bringing similar groups together by working on their difference from, and rivalry with, other physically or philosophically distinctive groups. The cruder empires assigned their citizenship by skin colour, the cleverer by skill or accomplishment. Either way, a more comfortable existence for most of the world was achieved by worsening deprivation at its margins. Global statecraft always stopped climbing a few steps short of the one-world summit, breeding resentments that in time caused the rickety edifice to crumble from the edges.

## Don't Take Me to Your Leader

Globalization is now learning to skirt these twin traps. It escapes personalization by swapping individual for institutional force. When processes, rather than people, are the driving force behind events, death cannot derail them. Today's globalization also maximizes coverage by dropping its old exclusive tag. An identity forged around what people are for, not what they're against, creates common ground not ringed by disaffected exiles. All can, at least in theory, march to the universal tune, with shared beliefs and values transcending particular views and opinions.

When unpopular change is needed, it pays to put the blame on someone else. But living culprits have a habit of biting back. Impersonal forces can be held responsible, where identifiable faces might protest that it was none of their doing. So 'Blame Someone Else' transmutes to 'Blame Something Else'. Opprobrium descends on the Gulag guard who was 'only obeying orders', but not on the politician who cuts social services because public debt constraints compel it, or the boss who orders dismissals because falling markets give the firm no other way to survive.

Decision-makers welcome the discovery of constraints that can be said to have tied their hands behind their back. In the search for such constraints, globalization fits a double bill. It offers a set of highly visible, yet strangely inscrutable, institutions of 'global governance', including a World Trade Organization to force open its trade channels, a World Bank to keep hold of its purse strings, an International Monetary Fund to arrange its emergency loans, and a Group of Seven to tell world leaders what agenda to set for the domestic cabinet table. These are the agencies whose logos get defaced, and whose conventions draw protesters' missiles, when anti-globalizers look for a visible target. But their leaders are too uncharismatic to become convincing hate-figures, their reactions to crisis too slow to mark them out as the architects of disaster. Real influence lies behind these institutions, in the complex of market forces and geo-political realities that drove their formation and determine their function. This influence grows when symptoms of the new development – rising flow of items, ideas and individuals across borders – are mistaken for causes, allowing an event to be ascribed to what are actually its effects. So globalization is as useful to wielders of power who claim that it cannot be stopped as to the wavers of placards insisting that it must.

## Buzzword Bingo

Awards and indulgent headlines are the start of a glittering career in showbusiness, and the beginning of the end for any other type of business. Being tipped for the top is the surest route to swift descent. Few companies are free from trouble once given the 'excellence' label. Second novels are routinely cursed by the weight of praise heaped on the first. The career of buzzwords like 'globalization' follows a similar pattern. They lose their sparkle just as they start to trip off the tongue. To get to the root of current controversies over globalization, it is necessary to start by digging beneath it, to the ruins of earlier grand designs on which it tries to build.

'Convergence' was the geo-political watchword of the 1960s. Then, at least to the growing number of state-employed commentators, academics and professionals, American-based 'capitalism' and Russian-led 'communism' seemed set to meet on the middle ground of mixed industrial economy and generous social provision. But even as political theorists were polishing this 'post-ideological' picture, frustrated Marxists in the East were admitting that central planning couldn't work. And free-marketeers in the West were arguing that they could no longer afford public industries or welfare states, which could survive only through a hand-over to private ownership and management.

After Europe and America brushed with stagflation (stagnant growth laced with rising price inflation) in the 1970s, the marketeers swept to power, with a mandate from welfare-weary electorates to dismantle the state apparatus. 'Market liberalism' became the economic credo of the 1980s and 90s, as voters were reminded that with every helping hand the Treasury sends a tax demand – and were persuaded that 'public goods' were no longer doing the public any good. Businesses that once looked to the state to build their roads, phone lines and power supplies with taxpayers' money, and seek rescue from the public purse when times got tough, were told to build their own infrastructures and raid their own contingency funds. Households that had come to expect the state to give them education, healthcare, pensions, unemployment benefit and help with housing costs, got a similar message to stand on their own two feet or look to charity for relief. Free markets were trusted to channel resources to those who could make the most of them, and to give the products to those who valued them most. And freeing prices and wages from state interference, so that they could adjust until every product found a buyer, was expected to end the bad old cycles of unemployment and inflation as demand swung above or below the available supply.

The promise of free enterprise faded as victims made others see the hollowness of their victory. In countries that went furthest down the free market route, businesses themselves were often the strongest lobbyists for more company regulation, publicly financed infrastructure, trade restrictions and state intervention to steer the economy away from boom and slump. This didn't surprise the free market's deeper-probing analysts, who had long ago realized that individual transaction in unregulated markets can never provide all of the goods and services required by a modern economy, ensure the balance of demand and supply that will keep the economy stable, or promote the innovation and investment that enables the economy to grow.

The so-called 'Pacific Century' made its appearance in the 1980s. Japan's return to commercial superstatus within 40 years of World War II seemed set to be repeated by South Korea's inside 20 years, and China's in less than ten. Where East Asia's tigers led, Indo-China's lions promised to stay only one step behind. Spurred by the need to fashion an economy from unfertile lands and scarce natural resources, these 'new industrializers' worked hard, saved harder, and assiduously imported and improved other regions' technology, until they could sell enough abroad to pay for the materials and fuel required for a decent living at home. Politically, their Confucian or Samurai inheritances seemed to have turned government into an instrument which fuelled economic expansion while curbing its own: stabilizing the economy, refereeing and redistributing private wealth creation for the common good, and building education and welfare infrastructures which market traders needed but could not provide themselves. Culturally, Asia could adapt modern technology to traditional family structures and artistic and religious sensibilities, avoiding the selfishness and destructive competition that threatened to drive Western powers from moral into material bankruptcy.

Yet in the 1980s, just as 'Asian values' of thrift and fair distribution were hailed as the unique route to economically sustainable and socially cohesive development, Japan's economy stopped growing, China's democracy movement wilted (amid accusations that its 10% growth rates were as bogus as its Communist Party election triumphs), and other new industrializers stumbled over trade imbalances and rising debt. The tigers ended the century licking their weakened-currency wounds, with their smaller followers having to turn back to the US for their economic lead.

The 'American Century', which had looked to be over in the

20th, then made its pitch for an extension to the 21st. Latin America was its usual mix of stop-go economies and come-and-go democracies; but to the North, in the USA, those who claimed to have knocked out the communist planners and outpointed the Asian tigers were happy to step into the consequent vacuum. From 1989, the US found itself without a geo-political rival as the Soviet Union fell apart, killed not least by its inability to match the Pentagon's war machine without crippling its civilian production. From 1991, the US enjoyed an equally unprecedented financial pre-eminence, as its stock market boomed – multiplying global investors' assessment of its economic wealth – and the dollar went worldwide as the currency in which all countries traded, borrowed and saved; the dollar also became the currency used by an increasing number for their everyday domestic transactions. The financial boom entered a uniquely virtuous circle with a roaring real economy, which saw US gross domestic product (GDP) grow for eight straight years, and unemployment fall to an historic low without the inflationary backlash that had stifled all previous upturns. Shares in American companies were being aggressively marked up because the US held the keys to the next century's major new technologies (micro- and opto-electronics, genetic engineering, software, networking, robotics) and had a method of organizing them – the market-disciplined, shareholder-driven corporation – which ensured that they would keep the advantage even when rival nations had acquired the good ideas.

The surge in share portfolios and other asset values helped, in turn, to reduce the cost of investment funds, so those corporations could outspend all others in getting ahead with the new technologies. Fast growth also reduced the US government's welfare obligations as pay and employment rose, enabling tax cuts that deepened private-sector pockets and reduced their capital costs still further. The same rise in financial wealth that gave companies more resources to produce things also gave households more scope to consume things, so that this expansion balanced and sustained itself long past the point where others had stumbled on empty shelves or unsold stock.

But even before America's growth machine ground to a halt in 2000, with its production stalling and share prices deflating, its 'New Economy' was showing tell-tale symptoms of its old disease. A yawning excess of consumption over production (hence of imports over exports), and an inability to save despite enjoying the world's highest income, had made the US perilously dependent for its investment on other nations' capital. This was imported on promises of rates of return which, by century's end, had moved from commanding height to American-pie-in-the-sky. US investment of that capital in new technology had done little to lift a slow productivity growth trend, its vast 'dot-com' outlays scrapped with one zap of a slow-to-open website. Its licence to print the world's currency (and borrow it back at zero interest) was at risk of being withdrawn, as Europe's euro and Asia's resurgent exports after the collapse of the tigers threatened to down the mighty dollar. Over-exposed on the world's financial markets as government and households borrowed to fend off recession, and over-extended in its military hot-spots by sole-superpower policing obligations, America was looking again like the classic sclerotic empire whose greatest height precedes its most precipitous fall.

## Rude Awakening

With America, Europe and Japan all falling short, no one could claim a monopoly on virtuous economic management or social design. It didn't matter if others came close. Small countries trying to spread their approach feared finding out that it didn't work on a larger scale. Large countries risked the charge of bludgeoning others into their mould. Globalization stepped in to fill the vacuum left by despair with one-nation ways. Where no one country had got everything right, mixing their models might just create the best of all possible worlds.

Like the big ideas it replaced, globalization was under way

long before it was in the news. Despite frequent setbacks and sharply different speeds, the world has been deepening its trade, financial and knowledge-sharing links for more than half a century. Until recently it was a slow and invisible process, noticed only by the immediate winners and losers. High-flying managers found their salaries spiralling as head-hunters took their battle across borders. Low-skilled employees saw their work dry up as cut-price counterparts did it for less. Between these exposed extremes, most saw the growth of foreign names on football score-sheets and foreign languages on packaging, but didn't connect it with any wider movement that might brighten or burden their own lives.

The acceleration of a trend, not the trend itself, is what brings long-term lifestyle changes under the spotlight. A smooth ride can inure us to the scale and speed of progress, but we feel it when someone prods the gas pedal or steps on the brakes. So it is the greater recent pace of global integration that has brought it to public and policy-makers' attention. The 1990s featured a picking up of output and trade growth among the world's major economies, the breaking down of their main ideological divide (between capitalism and communism), and the quickening spread of forms of democracy previously confined to a small élite of rich and stable nations. These developments were driven by, and added fuel to, the linking and lengthening of communication lines between countries. Their dialogue down these lines was made faster by common language and technology, and ranged more widely through shared recognition of planetary problems that required a united attack.

As a result, the new millennium began with a tide of optimistic future visions, in contrast to the spirit of doom that dogged past approaches to that ominous dateline. An arbitrary judgement of the Western calendar was eagerly added to the festivals of others, as all-night parties found favour even where political parties fear to tread. The discovery that computers, on which much of the new networking relied, were poised to commit collective suicide as the clock turned only added to the fun. The 'Y2K bug' looked like the ultimate disaster scenario, in which the disco lights would die just as the corks started popping, and falling airliners would drown out the fireworks as their auto-pilots travelled back in time. It turned out to be the first world-scale practical joke; and common mockery of computer-makers was added to the growing list of issues on which once disparate cultures could laugh with one voice.

Two years later, that laughter had given way to fear, as sudden slowdown threatened to throw the 'G forces' into reverse. The long economic boom, marked and perhaps reinforced by unprecedented commercial and financial de-regulation, officially ended with America's fall into recession in March 2001. A month later, the image of a new international unity, free of old ideological divisions, was shattered as the US and China squared up over a downed American spy-plane. The Palestinian–Israeli peace process, another test of the new world's conflict resolution, began to unravel again that summer. And in September, international terrorism staged its biggest show in front of the world's TV cameras, as hijacked planes felled the twin pillars of American commerce and defence.

## The Clash of Globalizations

For a moment, the '9-11' attacks on the World Trade Center and the Pentagon seemed to have revived the very process that they set out to destroy. Like climbers stopped by an avalanche within sight of the summit, the world was challenged to pull together for an even stronger assault. But when it was launched, against Afghanistan in October 2001, that assault showed a world already sliding back to its old, divided patterns. The Anglo–American alliance that spoke for 'freedom-loving peoples' against an 'axis of evil' stirred painful memories of earlier pacts, from the Crusades of the 12th century to the Cold War of the 20th. Britain tended to look westwards for its 'special relationship' while its European Union (EU) counterparts sought a shift of focus eastwards. Unable to live as one, the world always seemed to need some fundamental issue to divide over.

Debate over the best type of economy had swung, with communism's collapse, in favour of capitalist free markets. But new disputes reflecting different world views were not slow in erupting. What religious faith to believe in, who to trust with weapons of mass destruction, how to protect the environment, where to draw the line between private freedoms and public duties, whether to let scientists re-engineer life or trademark the results; these and many other divisions of opinion, long bubbling below the surface, found their release in the dusty rubble of New York's twin towers.

'Ground Zero', scorched launch site of the self-declared war on terrorism, reminded its intended victims that there were gods other than the market, and creeds beyond enlightened self-interest under a property-owning democracy. 'Camp X-Ray', the US colony in Cuba where alleged accessories to the atrocity were taken after their leaders fell, showed that legal rights and civic values were not entirely universal. Not everyone in the new world order was entitled to receive them, and the commanders of that order were not always obliged to give them out.

While the vultures circled these new bones of contention, globalization's core values were coming apart. The rising tide of affluence that had promised to wash away differences of geography, history and scripture had been halted by new barriers to economic growth. Against economists' predictions, free trade and capital movement widen the inequality of incomes and life chances, within nations as well as between them. This forces states to block market forces which adversely affect a powerful minority, even when their free play would be good for all the rest. Among the moves by US President George W. Bush that hit hardest against the present globalization, a tax on steel to stop cheap imports from doing harm at home ranks at least as high as the proposed attacks on Iraq to stop its military excesses from wreaking havoc abroad.

The possibility of ever-rising output has been challenged by excesses that depleted the natural environment; its desirability has been undermined by inequalities that poisoned the social environment. Alarmed at what unfamiliar cultures might do with their hard-won freedoms, old imperial powers revive their colonial ambitions under the guise of intervention to rebuild 'failed' states. Past empires fell apart when the colonialists' call to stay loyal for mutual prosperity was drowned out by their captives' insistence on sovereignty, even at the cost of initial economic dislocation. The new world, though forged around many market-places instead of one big shopping centre, has run into the same old trap. Global economy's losers complain of still having no clean clothes to wear, while the winners fear being left with no clean air to breathe.

## Rebels in Reverse

Globalization shaped around Western economies and their multinational companies has been challenged, from within that privileged minority, for its apparent indifference towards the majority of mankind. Half a century into the latest attempt to integrate world economies and cultures for a better shared future, half of the planet's population still lives on less than $2 a day. Though large in number, these victims lack a voice. Their own governments, forced to look abroad for financial and logistical support, are frequently indifferent to their complaints. So their words of dissent are increasingly spoken for them by 'anti-globalization' campaigners based in the rich world, whose record of honourably disrupted superpower summitry already runs from Seattle in 2000 to Genoa in 2001 and Washington in 2002.

The protesters oppose a 'globalization' that has become the new shorthand for forces breaking down commercial protection and cultural insulation. They propose an 'anti-globalization' that crystallizes complaint against the world's new interconnectedness, uniting those who acknowledge its potential uses but who see something nasty in the watershed. Their 'movement of movements' has gained worldwide attention because it fills a void left by previous movements when they shed their ideology, lost sight of their common enemy, and expended their remaining energy on fighting among themselves.

After their unexpected triumphs against apartheid, the nuclear arms race and the Vietnam War, rich-world rebels had ended the millennium without a cause. Hopes for an alternative world order had been crushed by communism's retreat to the poorer ends of Cuba, Korea and Camden. Morale in more human-centred business structures burned low as cooperatives squeezed their payrolls under creditor pressure, and mutual societies went public with a mission to raise 'shareholder value'. The non-profit-making 'third sector', once carefully perched between public service and private profit, was co-opted to plug holes in a dwindling welfare state – forcing the street fighters to scorn the soup kitchen and brush aside the collecting tin on the grounds that such charity merely helped the government dodge its social obligations.

Stripped of their traditional solutions, protesters seized eagerly on a slogan that gave them the whole world to campaign against – without the inconvenience of having to explain what they were actually for. Embracing anti-globalism showed that capitalism's critics could think beyond their national borders. Their mission was to resolve the market economy's problems, not just displace them across borders. Nationally-based socialists and left-liberals were accused of defusing the Western class war by shifting exploitation and alienation to developing-country sweatshops. Marx's famous call to action, long blunted by disagreement on what 'Workers' were, could now be refloated with the emphasis on 'Of The World, Unite'.

Anti-globalism also aligned left-leaning opinion with a previously circumspect environmental movement. The Greens had likewise accused old territorial-based socialism of munching all the world's resources just to fatten a small proportion of its people. The eco-warriors tapped a rich seam of colourful protest tactics, often borrowed from the surprisingly well-organized Anarchists, and reinvigorated by the computer-age anarchy that seemed to flourish on the World Wide Web. Common cause was even found with a world-weary feminist movement, which had detected a heavy male hand behind most previous protest banners. Women perform more than half of the world's paid work and almost all unpaid work – so when victims strike back against this new division of labour, it's often a feminist force that downs tools and goes on the march.

Once the fight got under way to save the world's underprivileged from a web spun by its wealthiest, only one group proved reluctant to join in. These were the world's underprivileged themselves – many of whom had signed up to the new multinational networks as customers, suppliers and employees, and resented attempts to dismantle or disable those networks. Those in whose name the world was stopped were peculiarly reluctant to recognize the efforts being made on their behalf. It was a familiar feeling to anti-globalizers reared on anti-capitalist campaigns within one nation, whose strongest and most uncomprehending critics had always been the workers whose liberation from employment they were trying to secure.

Developing-country leaders in the 1990s signed their free-trade treaties, approved their multinational investor protection codes, and then stepped outside the summit hall to screams of abuse, unable to understand why such wrath was raining down on them. To those out to derail such gatherings, these were the world's incestuous élite, finding a common taste for conference dinners and forgetting the common suffering of the billions they represented. Or they were well-meaning reformers who could do no right because international economic forces had tied their hands. Whichever way, because their own associations and aspirations were so thoroughly international, these negotiating élites and their national political machines were seen far more as problems than solutions.

National governments were seen as incapable of creating a

fairer world, supra-national structures like the World Trade Organization even more so, because they ranged too widely to represent all the people beneath them, and rose too high to be receptive to those people's needs. Anti-globalizers sought their North–South links not between the national governments of rich and poor, but between non-governmental organizations (NGOs). Campaign groups, community organizations, fair-trade cooperatives, credit unions and other self-help networks could escape the institutionalized inertia of old top-down protest movements. And they could counter the obsolescent national political hierarchies by forging horizontal links directly between their patches of common ground.

But worldwide reconstruction will continue to elude the new movement, until the contrarians acknowledge some basic contradictions.

## False Negatives

Few of the diggers and dismantlers oppose globalization in all its forms. With their tendency to leave the urban cabbage patch for a cheap flight to the next stormy summit, or a visiting professorship, it would be hard for these well-dressed dissidents to do so. Their biggest missiles are reserved for a globalization orchestrated by multinational companies, built on a free trade which locks in early industrializers' advantages, and extended to natural and cultural products which defy commercial logic. Blinded by the dust they raise, they often fail to see that the doors they push against are already open. Some are off the latch because pro-globalizers have already conceded their argument, some off their hinges because the biggest battle has already moved on.

Opposites attract. So when globalization strikes, localization is the antidote most instinctively reached for. It has the weight of history behind it. The world was hard to handle until chopped up into manageable sections. Politics leans strongly on the side of the local, because small territorial units are hardest to dictate to, and easiest to govern by consensus. Politicians who acquire immunity to the boldest stares of Larry King or Jeremy Paxman still wilt when forced to look electorates directly in the eye.

Those sceptical of 'small is beautiful' in government can point to the embarrassingly low turnouts in many local elections, and town halls' extra proneness to commissions, bungs and bribes. They sneer at the origin of many small states as bigger states that fell apart, because people were too incompetent to make big economies and administrations work, or too xenophobic to share politics and culture with those of different appearance and religious convictions. But localizers countercharge that it is central government's commandeering of powers from the local level that makes voters lose interest in borough and county polls. They can use the even lower turnouts in European Parliament elections to show that apathy gets even stronger when power climbs further above ground level. They can hail the fragmentation of large states as a restoration of the autonomy that megalomaniac state-builders had previously trampled on. And they can point to local anti-corruption clampdowns leaving national politicians and bureaucrats to be the more frequent and extravagant graspers of greasy palms.

Under economists' theories of free trade (to be reviewed in chapter two), division of territory supports multiplication of wealth. The smallest territorial units tend to be the biggest gainers from trade, making the largest productivity rise in operations they choose to retain, and so enjoying the widest range of newly-opened import possibilities. Singapore, Hong Kong, Monaco, Ireland and Slovenia were regional growth stars in the 1980s and 90s, and many more 'micro-states' are tipped for the top in the 'Naughties' and beyond. The smallest states are generally the best at solving the 'public goods' problem, because they are where the destination and effectiveness of public funds is easiest to see, and so collective action easiest to agree. They are the best at adjusting to sudden upheavals in the global economy, through circulating their own currency and letting its exchange rate float up or down against others who weather the storm in different ways.

## Faulty Alternatives

In viewing localization as the remedy for globalization, critics overlook the very powerful forces that rage in between the two. Both vertical extremes have an intermediate foil in the form of regionalism – the pursuit of economic and political integration above the national but below the world level, which we will look at more fully in chapter two. When economic power passes downwards from the state towards private enterprise, political power passes upwards – but not to the global level. It tends to get trapped beneath opaque legitimacy ceilings and external tariff walls. The power that rises to regional level is often, as anti-globalizers allege, closely allied to the power that drops to enterprise level, helping big business to pursue its profit-making 'free market' agenda.

The smallest states have no assurance of success. Some are restricted by a lack of the critical numbers of people, like-minded groups, production clusters and commercial connections needed for economic efficiency and social stability. Many of the micro-states that have grown rich did it by exploiting tax or regulatory loopholes in a beggar-thy-bigger-neighbour fashion: enriching the life of their thousands at the expense of the many millions who lose their welfare-building tax base as moveable wealth and income hasten to the tax haven.

Any power that transcends the region and extends across the world has a contrasting, countervailing effect on corporate interests. So in attacking institutions of 'global governance', localizers are cursing the very gods from whom they should be seeking help. Those acronyms which anti-globalizers parade as universal problems – IMF, WTO, OECD, G7/G8 – are, in practice, the closest they have to world-beating solutions. If protesters' complaints are valid, Earth risks biting the dust not because these multilateral agencies are too powerful, but because they can't keep nations and regions on a tight enough rein.

An unequal division has allowed some nations to wield economic, cultural and colonial influence over others. The European language and Californian accent with which globalization's advocates speak, and the Japanese-designed electronics through which their words are relayed, reinforce the rebels' view that 'just one world' can be 'one just world' only if we cease to try to unify it on the biggest hitters' terms. But if attempts at world *political* consensus are abandoned, the way is opened to unification on purely *economic* terms. With the market-place chosen as humankind's common ground, geo-political power goes to those with most purchasing power. Global political institutions, by contrast, assigning votes in proportion to the number of mouths their members have to feed rather than the national incomes of their sponsors, may be the best hope of swapping the present bias towards the buyers, for one in favour of those with least to sell.

Personal interest has crossed wires with local concern in shaping the anti-globalizers' agenda. Getting back down to Earth, in the sense of seeking to rebuild the world from its local roots, offers two further great attractions over previous calls to arms. It means that the protesters can do something practical on the ground. Waving red flags on a big protest march has rarely set the winds of change in motion, and signing up to a big, bureaucratic anti-poverty charity makes the subscriber poorer without necessarily making the developing world any richer. But campaigning for a new pedestrian crossing or digging your own cabbage patch are practical contributions which, if done widely or seriously enough, really could change things. Anti-globalizers can also have great fun preparing the ground. The world may face a crisis of calorific and humanitarian value, but (protesters say, because) there is no shortage of tempting Exxon billboards to deface, and McDonald's shop-fronts to dismantle, as the grass-roots reconstruction gets under way.

## The U-sual Suspects

Technophiles are fond of telling us that we can't turn the clock back. New ideas are here to stay. And as we can't put gadgetry's genie back in the bottle, the only way to correct its early blunders is to let it continue. Cars may poison the air, but if we'd stuck with horse-drawn transport, the weight of waste would have stifled us even before it buried us alive. Oil and gas may warm the planet, but if we hadn't started burning them, every forest would be firewood by now. Technology's side-effects can usually be cured – but only if the main event is allowed to continue. Turning back because of teething troubles, say the technophiles, means swallowing the rind without tasting the fruit inside. If an innovation hurts at first, the lesson is to refine it until it gets comfortable, not renounce it with only the crash-test dummy's injuries to show for the excursion.

Unlike most technologies, a society can easily go backwards. We can't disinvent the nuclear bomb or the transplanted gene, but we could soon lose the laws and civic values that let us control these for the common good. The existence of a reverse gear is unsettling when present and future look good, but reassuring when the path ahead is invisible or inhospitable. The world's politics are today so turbulent, its economy so unstable, that our biggest danger is in staying where we are. If we can't or won't move forward, then the only safe ground is behind us. This is missed by the many protesters whose remedy for a runaway world is to stop it now, and let the victims climb straight out.

A backward or forward step is needed, because simply stopping things now risks freezing history at its most inhospitable moment. If globalization were to stop right now, it would institutionalize the present unfair distribution of privilege and poverty. Parking near the bottom of the U-curve means getting trapped where the environmental impacts are unsustainable, and the economic gains unjustly spread. Breaking up the bandwagon because too many people jumped on may seem perverse, but it may make even less sense to stop it in its tracks while it's sinking into the mud.

Economists, whose work often features in what follows because they have analysed and advocated globalism almost from the start, see much of world history in cycles or U-shaped curves. Their prevalence (along with supply and demand 'curves' whose crossing reveals market-clearing price and quantity) shows how extensively globalization analysts make use of models – and how, as the models get ever more scantily clad, it becomes increasingly important that they show the world the rosiest set of curves. U-shaped curves are a technical twist on the ancient wisdom that 'things get worse before they get better'. They are the statistical heart of many optimistic pronouncements on economic and environmental trends.

The classic U is like the upper half of a wine glass viewed from the side. Its usual purpose is to show that the half-empty wine bottle is actually half-full. The U's enthusiasts can take a seemingly inexorable decline, show that the deterioration is slowing, and infer that in a little while it will transmute into an inexorable rise. Be patient, they plead, and all lost pleasures will be retrieved, even exceeded. Arresting a process because its initial effects are bad, they say, is like jumping off a rollercoaster because of giddiness half way round: just occasionally, such evacuation can avoid a spectacular crash, but more often, it means piling on the early pain without feeling the later payoff.

The global financial and commercial systems on which the current world order has been built can, at times, beat any rollercoaster for discomfort. The solution is not to leap off, or to stop the systems in their tracks, but to straighten and safety-test those tracks. That task is likely to require more globalization, not less, and failure to recognize it could leave the anti-globalizing movement stuck at the low point of its own U-shaped trajectory.

## So What's Coming Up?

This book tries to explain why the rebels are right to see globalization as important – and almost always wrong in their diagnosis of what's bad about it, and their prescriptions for what should be done. Recognizing that today's explosion of globalization, and the protests of its malcontents, are the result of a long, silent build-up, the book starts by tracing this process back. The economic forces and theories at work turn out to have been around for at least the past two centuries, and the recurring political and social features to have been visible as far back as ten centuries. Then, acknowledging that the story is being told in a new way, and has acquired some previously unseen episodes, the second half of the book looks forward to where globalization is likely to end up if not corrected; and to where its opponents would like it to go, if their often inverted reasoning goes similarly unchecked.

Chapter one examines the globalization critics' favourite target, multinational companies, and argues that their overseas march is less an aggressive effort to colonize the world than a defensive tactic for holding on to vulnerable business at home. Far from being the builders of a ruthless world market, multinationals are constantly trying to escape it, and it has been governments that have promoted international integration in an effort to curb big corporations' market power.

Chapter two assesses the traditional economic basis for globalization, arising from international trade, whose proposed all-round gains are the subject of much preaching by politicians and even more creative accounting by mainstream economists. In extending the effects of a division of labour *within* nations, which generally raises everyone's income, to a division of labour *between* nations, standard theories make several heroic assumptions about competitiveness of markets, income redistribution arrangements, availability of technology and the dynamism of chosen specialist areas. It is not the sort of heroism that wins medals, apart from the occasional Nobel Prize. (And even then, not all of Nobel's descendants think economists worthy of his accolade, some complaining that the subject scarcely qualifies as a science and is certainly not of a type to improve the human condition.) As far as economic thinking affects today's political leadership, it does so in the direction not of a world-unifying globalism, but of a deeply divisive regionalism. Global and local turn out to be less polar opposites than partners, with the over-reaching region coming between them.

Chapter three finds that international capital movement is potentially a far more powerful global economic engine than international trade, but one with a tendency to run in reverse, embarrassing advocates and opponents alike. The problem with capital is that, having accumulated an abundance of the stuff, the more advanced economies have split it into many different forms. Poor countries that import some versions can't make them work, because other versions refuse to come in to support them. More recently, even the exportable varieties have shown an awkward tendency to flow from poor to rich, reversing the direction of travel that economists predicted, and on which gains from world capital movement were predicated.

Chapter four considers where a world economy driven by free trade and capital movement is heading, and sees an impending abyss. The system has a perennial difficulty in consuming all it produces, which it tries to solve by preparing to produce more. It can escape the economic problem of too little consumption only by running into the environmental problem of too much production. Anti-globalizers can rightly accuse the 'free' market economy of both taking too much from the planet and of giving too little to the people. So the chances are high of making one, if not both, of the charges stick.

Chapter five looks at the role of government in creating or solving this impending crisis. Far from being muzzled by multinational capital, as many anti-globalizers claim, the state turns out to be the architect of an integrated world economy. Taking the trade barriers and capital controls away was an attempt to force the multinationals back under some form of competitive discipline. The attempt has backfired. But the economic imbalances created, now transposed to global level, only magnify government's role in stabilizing the economy's demand side (the total amount that customers buy) and regulating its supply side (the quantity and quality of goods and services that companies supply). Globalization means that national governments must, in turn, be coordinated and regulated by higher authority – most likely to be provided by the G8, WTO and other multilateral institutions, once the anti-globalizers realize them to be the nation state's salvation rather than its subversion.

Chapter six leaves the economic realm to examine the interplay of arts, culture, and technology with globalization. It finds commercialism penetrating these areas, with varying degrees of success. But culture and knowledge production, differently structured processes with differently behaved products, also have the potential for pulling commerce onto a more sustainable course. The 'knowledge economy' modifies the market economy through channels that the market economy could not have created, and what flows down those channels could either make globalization wobble or make it work. In seeking the first use, protesters risk ignoring the potential of the second.

Chapter seven travels to America, to discover whether the last surviving superpower is using globalization as a disguise for its own world domination, or merely trying to hold its own in a world that's still beyond it. While most of the planet now lies within America's economic and military grasp, most of it still defies Americans' mental grasp. What looks like an attempt to colonize the world is simply an attempt to make it comprehensible and to cool down the culture shock when they step outside – which, because of their commitment to being the world's champion consumers, other nations must force them to do, at least until they find a more multinational way to fend off the under-consumption crisis.

Although much current globalization is thinly-veiled Americanization, today's planetary pooling goes at least some way towards transcending the USA. A much more even spread of benefits is needed to make the process acceptable to all nations, for reasons and through mechanisms that this book will try to explain. But even where such balancing means trimming some of America's distributional privileges, the effects can be counter-productive. After the calamity of 9-11, with the sole superpower's stock markets sunk by nano-profit technology and its shopping malls haunted by the calling-in of credit card debt, other nations might have welcomed a release from the wounded giant's shadow. Instead, they found themselves blinking in a harsh new light. Fear of a rampant America taking over the world had turned to fear of a rueful America stopping it and getting off.

The States may be a bloated balloon which the rest of the world has tried to burst, but if the two sides disengage then both will come to grief. US withdrawal symptoms were already evident before 9-11, from unsigned Kyoto environmental protocols to unpaid UN dues. Americans ultimately step outside their oversized home only to fix the power supply and check the plumbing. They often siphon off neighbours' resources and trample onlookers' toes while doing so, but surrounding streets would suffer much more if they put up the shutters.

Stepping beyond its boundaries in order to secure them has covered a multitude of American sins. But the American wish to reconfigure the world so as not to have to leave home is one from which the world can only gain – by going along with it. A planet that did not recognize, and seek to share, America's substitution of trade for migration would be a world swapping exchange's uneven pull for self-sufficiency's unstable push – only globetrotting because there's something to run away from, only border-hopping because frontier fences have climbed too high to stride.

## Taking on the World

Whether hailed or hated, globalization is not new. It is the way things always were, before territorial division and tribal loyalties cut in. It is a state that the world continually moves back towards, only to be deflected by military flare-ups and monetary downturns. There was only one world when ancient cosmologies and creation myths were still in production. But localization got under way early, if the Tower of Babel and various Founding Father tales are still to be believed. It was only a century ago that ways were found to transmit voices around the world, only 50 years since people could buy tickets to fly around it, and only five years since most could offer the world their opinions, biographical details, get-rich-quick gambling and share tips or downloadable joke supply through the electronic post.

Much of the current anti-global outburst is an attempt to cling on to past localization, when stronger cross-border links are a better way forward. The BBC World Service scared longer-distance listeners for many years by announcing the conclusion of each bulletin as 'The End of the World News'. BBC national radio made amends with a daily lunchtime programme called 'The World at One'. But it was often to the champion wife-throwers of Estonia, or world-touring nose flautists from Outer Mongolia, that the domestic airwaves looked for their upbeat final story. While unity can sometimes be achieved through a universal sense of other nations' eccentricity, globalized observation is still very far from shared interpretation when it comes to harder news.

This book is designed to dispel any fears that 9-11, the Battle of Seattle or the Teletubbies' conquest of China is the end of the world as we know it – if only by suggesting that we still don't really know the world that well; and that what we know is not what is claimed by those who forecast its demise.

CHAPTER ONE

# Bad Company

*Most critics of globalization are not opposed to all of its manifestations – only the present dominant form, led by profit-driven multinational companies (MNCs). Foes portray today's big business as exerting worldwide economic and political power far more effectively than past imperial rulers. But far from being the product of runaway market forces, MNCs suppress these, owing their success to a privately perfected form of communist central planning. The governments alleged to have been swept aside by corporate expansionism are actually those which unleashed it in their own defence.*

You can't escape the world any more. Cross-border links lurk wherever you look. Travellers' tales that once hinged on how different foreign lands were now revolve around how similar they're becoming. The ends of the Earth aren't far enough away to escape McDonald's golden arches, Ford's blue oval, Benetton's united colours or Nike's swoosh. If you ever find a bar not serving Heineken or a car not powered by Shell, you're either in a Disney theme park or under the influence of Monsanto's more exotic GM herbs.

These footloose logos give globalization a household name. But much of the breaking down of borders is taking place outside direct corporate clutches. The critics of multinational companies like to present them as the architects of a shrunken and standardized world. Those companies' publicity departments tend to accept the claim, and take it as an accolade. But border-hopping businesses are less the cause of globalization than its consequence, and the process of one-world construction goes far beyond them. Although acclaiming, and condemning, MNCs as the humanized face of a unified planet, the critics of the corporations need to look elsewhere.

## Invisible Villains

A good fight needs a personal foe. Many worthy campaigns have flopped through disagreement on who to aim the missiles at. We need an image of almighty gods to comprehend what's really good, and twisted smiles on 'Wanted' posters to know who's really bad. Battle lines are more easily grasped when St George is given a dragon to slay, when Dorothy has a specific wicked witch to pursue down the Yellow Brick Road.

When, in late 2001, America took on the al-Qaeda terrorist network, its tangled web was not a clear enough target to aim the high-tech weapons at. One notorious member, Osama bin Laden, had to be identified as the hate figure around which all atrocity revolved. The fire-resistant beard came to symbolize, both for followers and foes, what the States' smart bombs were trying to lay to rest. This personification – and its dual usefulness to perpetrator and pursuer – followed a long tradition which saw Rome's foundation traced to Romulus and Remus, Russia's communist productivity miracle symbolized by Stekhanov, America's green revolution single-handedly sourced to Johnny Appleseed, and Britain's early industrial machine-breaking made the exclusive terrain of Ned Ludd.

New technology which amplifies individuals' power, and new laws which heighten their responsibility, have raised world demand for demonization. When a weather phenomenon called El Nino caused extensive wind and flood damage in the late 1990s, devastated North American householders were not content to fume at intangible climatic effects. They leafed through the phone book and discovered a man called Al Nino, to whom their irate and derogatory calls were then directed. Abstract nouns, like abstract art, are something the world has difficulty getting to grips with. Taking arms against a sea of troubles is a public relations nightmare, compared with the propaganda triumph of harpooning a single leviathan.

Except when you're fighting globalization. Then, the lack of a recognized skull on the crossbones is part of what you're objecting to, part of what makes the corporate flag worth burning. MNCs' worldwide march is made more scary, and less stoppable, by the fact that no one strategist seems to be behind it. The architects of corporate-driven globalism are not a sinister ruling clique. They are a disparate band of bosses whose enterprises are ostensibly in cut-throat competition. While a few set up outposts on every street, and put themselves at the head of the parade, most prefer to be anonymous cogs in an invisible machine.

The grain on which most of the world relies for its subsistence is traded by just three companies, whose private owners are rarely named and whose financial dealings stay completely under wraps. Air travel is run by a bevy of four-letter acronyms, from IATA to SIDA. Its stay-at-home alternative, the Internet, prefers longer acronyms, but this does little to blow the cover of those who assign the dot-com domain names (ICANN), or who work to harmonize state-specific laws (NCCUSL). The lower the profile of an international organization, the longer the strings it turns out to be pulling. Small wonder that 'Silent Takeover', 'Secret War' and 'Invisible Empire' are among the grudgingly respectful terms by which globalizers' enemies have chosen to know them.

The facelessness of big business as it rides into the global village, and its tracelessness when mystery deaths and sudden dustbowls follow, is one reason why rage against 'one world' is so easy to arouse. Cross-border forces are all the more sinister for being unseen. Even before its gaudy shop-fronts became the chief target for multinational-bashing, McDonald's was learning to play down its expatriate presence – encouraging the Japanese belief that the bulging bun was their own invention, calling its coffee-shop chain Aroma, and keeping quiet about becoming Pret A Manger's biggest shareholder. Having seen Big Macs forced to bend to local taste, and then overtaken by alternative snacks, the new tactics relied on not letting customers know what distant hand had stirred their coffee, or spooned their luxury spread.

## Accounting for Taste

Posing as home-spun players, and hiding their global roots under local branches, deepens the menace of the MNCs' disguise. A favourite jibe against the border-hopping business is that it rams the same product down every throat, driving out local varieties and denying the right to be different. Middle Easterners must blunt their taste-buds on freeze-dried, flavour-drained instant coffee; Asian viewers curl up in front of universal pictures fed through standard set-top boxes; Africans take to dirt tracks in cars designed for smooth American roads.

The anti-globalizers' doomsday scenario shows MNC sales forces and satellite broadcasts stamping uniform patterns of dress, speech, diet, politics, hairstyles and ball-games on the world, more effectively than any past imperial power. The image is of Ghanaian villagers forced to drink Coke because its bottling plant takes all the clean water, or Nepalese schoolchildren stumbling through the snow in Nike trainers (sewn by younger counterparts in Mexico until child labour campaigners kicked up a fuss). Even those who never go beyond their own corner shop find themselves fettered by an international supply chain. If you made a last stand for self-sufficiency by never leaving home again, survival would still depend on 'Products of More Than One Country' – delivered to your imported softwood door by couriers headquartered half a hemisphere away.

It is easy to mobilize wounded national pride when the preferences of shoppers in Madras are dictated by product developers in Madison Avenue. But in practice, very few of these 'power brands' straddle the globe in the way that their castigators claim. Marketing history is replete with success stories lost in transplantation: from washing machines that propagate stains when fed Indian water, and brand-name beers that outreached local tastes and incomes in China, to cars called Nova in Spain (where it translates as 'doesn't go') and MR2 in France (an early casualty of scatological phonetics). To forestall such cross-border flops, manufacturers now indulge local variations much more frequently than they try to stamp one taste across all territories. Even mighty Coca-Cola outsells its eponymous core product with canned tea in East Asia and less exotic fizzy fruit drinks in former communist Central Europe.

Cross-border variation is a logical extension of MNCs' product strategies at home, which involve honing the product for different segments within one national market. Car makers have long used variations in engine size, interior quality, paintwork and optional extras to pitch the same vehicle at different income and aspiration levels. World-building auto groups like Volkswagen have gone further, constructing different shapes – from cheap Skoda hatchback to luxury Audi sportster – on the same chassis, one piece of engineering primed for multiple journeys with widely spaced price tags. From phone users signed to different service grades to computer buyers selecting their chips, customers have grown accustomed to rounding off their choices with a personal touch.

Suppliers can allow this, without significantly raising costs or delaying delivery, because of the new technology and training trends. By confining differentiation to the final stages or outer edges of a product or service, businesses can mass-produce the core while retaining customization or personalization at the periphery. Easy reconfiguration of computerized machine tools lets them switch easily between specifications. Digital communication and just-in-time stock management helps clients fine-tune their own design, with production often starting only when orders and payments are received. Staff needed to manage the robot arms are also trained to repair and re-program them. Others are made more 'flexible' by having their more exacting tasks taken over by machine. The resultant mass customization has long offered a way to widen the gap between prices and costs in a particular national market. The move abroad can make it even more effective, because of the much better prospects for keeping buyers in one market physically separate from those in another.

Multinational 'price discrimination', setting distinct groups of customers a different charge for the same item, is a powerful way to squeeze out extra profit. At the extreme, charging everyone the most they are willing to pay would be vastly more profitable than setting one averaged 'market' price, which may be much less than the richest would accept while being much more than the poorest could afford. But discrimination breaks down if those charged lower prices are able to buy extra, and resell it to those being asked to pay a premium. Until trade barriers were removed and transport costs reduced, geographical distance between markets was enough to make such discrimination safe. Even if the price in one national market exceeded that in another by more than the gap imposed by tariffs and fuel, the route to easy profit by buying low in one and selling high in another was blocked by lack of knowledge about the opportunity, or instability in the two countries' exchange rates which made it too risky to exploit.

With the freeing of trade and stabilization of currencies, price discrimination on identical products has become harder to maintain. Sellers who price differently in each country are in peril from 'parallel imports', as customers go to the cheapest market or, more dangerously, large retailers switch their sourcing there so that cheap supplies leak into the lands where they were meant to stay expensive. Manufacturers have sometimes tried to stop these grey-market sales by legal means. In 2001, Levi's persuaded a UK court that a big store chain, Tesco, should not be allowed to buy its jeans abroad and sell them more cheaply than the recommended price at home – arguing that the cost-of-living gains to customers of paying a lower price would be offset by the lifestyle-quality loss, as their designer clothing was devalued by being sold alongside eggs and packaged bread. But parallel imports can be stopped far more effectively, without involving the courts, if each nation's product takes on a different personality to match its different price. Car makers have always used the British peculiarity of driving on the left and wanting steering wheels on the right to keep the UK market separate, and to charge higher prices than in the rest of Europe.

The typical multinational product is not a one-size-fits-all car like the Mondeo, but a beer brand like Budvar or Newcastle Brown: local town or village association up front, multinational parentage kept well into the background – and taste a crafty compromise between the unique idiosyncrasy of the small production run and the low price with constant quality made possible by the large. World business is being taken over by companies that think globally but transact locally. The slogan which anti-globalizers believed would undermine MNCs has become a strategy now lifting them to new controlling heights.

## Nameless and Shameless

Armed invasion at least has the merit of letting the victims know who's attacking them. Foreign forces' strange language, customs, battledress and tendency to run red lights with their tanks leaves little doubt as to who the intruders are. The unarmed, often invited presence of multinational companies lulls local people into an acknowledged and accepted form of occupation. Any doubt about succumbing to global corporate domination is dispelled when we no longer even notice that we're under its command.

The first foreign ventures, focused on mines, plantations and export-processing *maquiladoras*, were correctly regarded as plundering native resources to make profits abroad. While oil, topsoil, the clothes from workers' backs and the sweat of their brows are being stripped from the landscape for sale abroad, foreign-owned enterprise can rightly be seen as excuse-free exploitation. MNCs often claim, with some justice, that they create more jobs than they destroy by offering new input supplies and markets for local employers, whose pay and conditions they generally improve on. When – as in Singapore after 20 years of opening for business, and Hungary after less than ten – MNCs account for more than half of a small economy's total exports, they can present this as seizing an otherwise missed opportunity rather than posing a strategic threat. But these successes, even if achieved, are in a game fundamentally reshaped by the multinational presence. A country made 'export-oriented' by MNC presence sees its production redirected from domestic needs to foreign wants, and its natural resources consumed or combusted abroad, with only the waste gases wafting back to it. When Cubans wait for water off a truck while their foreign hotel guests can turn the taps, or Colombians go hungry beside ex-grainfields that now grow flowers or drugs for air-freight to the US, the 'open door' seems to have given the gate-crashers a much better time than the embattled party hosts.

But later, larger waves of foreign direct investment (FDI) have moved away from old colonial extractiveness, towards using local resources to meet local demands. Smaller countries' success at raising their income (whether because or in spite of MNC involvement), and larger countries' opening to trade with them, has turned an increasing number of them into sizeable markets in their own right, not just useful bases for export to other markets. This brings a new set of charges against international business: that it addicts buyers to its global brands before local alternatives can get established; or that in taking over and reshaping local brands for mass production, its market monoculture stamps out precisely that variety that made the foreign trip worthwhile.

Changing native names and labels, without revealing their

new non-local ownership, excites a special brand of anger when the national products in question were also formerly natural. MNCs have used intellectual property laws passed in their home country to make proprietary claims on what hosts regard as common property. Crops they have cultivated, and lived on, for centuries suddenly reappear under a trademark, with instructions to pay royalties and not re-plant the seeds. Genetic variations passed down through generations suddenly pass to a foreign parent company, which counts the decoded chromosome as its own. The privatization and enclosure of land, previously preserved as open fields and worked in common, aroused the first big backlash against capitalist values at the start of the industrial revolution in Europe. Present-day commercial seizure of what grows on and feeds off the land, as science moves on from researching to redesigning nature, re-ignites that righteous indignation as post-industrialization gets under way.

## Giant Killers

As befits a world-scale production, corporate-led globalization is now a drama of the biggest stars. An ongoing merger-and-acquisition wave, gathering pace since the late 1980s, has given rise to commercial empires of unprecedented size. In globally standardized products like aerospace, cars and computer software, two or three companies now tend to dominate the world market. Airbus is increasingly the only alternative to Boeing, KPMG and PricewaterhouseCoopers the only places to go for financial advice, the Mac the last redoubt against an all-pervasive PC. Even where local knowledge and adaptation win out over the efficiency of central coordination, small players are linking with the giants by joining their global alliances or franchising their global brands.

Large though they may have been on a national scale, even the biggest businesses were generally confined within national borders until 20 years ago. Coke, the world's iconic soft drink, still made almost two-thirds of its sales in the US until the mid-1980s, when its Western European units broke through the soon-to-fall Berlin Wall to sweep eastwards. In the next ten years, a push into emerging markets more than doubled its global customer base. Wal-Mart, the world's biggest retailer, waited even longer before stretching its supply chains outside America, only arriving in Western Europe in the early 1990s. No US company sold more than half of its output abroad in 1992; the only ones that did were based in European countries with smaller home markets. Ownership, via share and bond holdings, was even more concentrated in the country of origin.

Before 1980, when a company couldn't make or sell any more of its product in the home market, it usually looked around for other products to sell to the same domestic customers. PepsiCo bought into food-processing and restaurants, Coke acquired a film studio and a packaging company. The UK's biggest banks started doubling as insurers and mortgage lenders, and leading supermarket chains tried their hand at DIY. In Britain, the wisdom that good management could do anything – even opposite things – was exemplified by British American Tobacco, which processed its poisonous weed while running a chain of health food shops; and Tomkins, which at one time made both guns and the materials for cleaning up bullet-wounds. Asked by shareholders to explain what logical thread could possibly run through its apparently random spread of activities, the US conglomerate 3M eventually alighted on 'things that stick'. Luckily they didn't make trouser zips or gearboxes then.

Shareholders, who'd started to wonder if top managers were doing the best job for them, often weren't happy with answers like these. Under shareholder pressure, corporate strategy has shifted substantially in the past two decades, from diversification at home to focused expansion abroad. Most MNCs have shed their 'peripheral' operations at home, concentrating on the 'core' activities they're best at, and using the advantage to capture other people's markets. McDonald's went into the world as its expert in flipping hamburgers, and has been notably less successful at serving other fast food types, such as coffee and sandwiches. Nokia's fame came from pushing a mobile phone into all the world's pockets, rather than filling out its original forest-products portfolio by offering Scandinavians more tissues to wipe their pine furniture with. Whereas conglomerates like General Electric and IBM once dreamed of offering everything that the automated home or office wanted, today's MNCs make a virtue of their dedication to one distinct product or service. Toyota even calls its luxury car the Lexus, and Hewlett Packard hid its cut-price printer under the Apollo name, to disguise from customers that its brand stretched as far as down-market segments of the same equipment type.

Slimming down to core products, and diversifying across national borders instead of product lines, was not a move that multinational managers chose to make. It was generally forced on them by impatient outside shareholders and creditors, who believed that there was more profit to be made in exploiting current product strengths in new territories than in tackling less familiar areas on domestic turf. The move was enabled by a widespread relaxation of competition laws, removing restrictions on domestic market share (so allowing more investment in the same line of business) because there was now an international market in which the national champions, even when monopolies, were still relatively small.

The massive size of these new combinations mocks the claim that we're still living in a market economy, at least as defined by market economists. Markets only work – allocating resources efficiently, expanding them rapidly and making innovative use of them – if there's free competition and free flow of information. The mega-corporation stifles competition, by reducing the field to a handful of big buyers and sellers. Though they stop short of holding a monopoly or formally colluding with other big names, suppliers have serious market power in this 'oligopoly' situation. By studying each other's costs and following their price signals, they can mimic monopoly in setting a price above the market level, raising their profits while restricting the number who can afford the product. By keeping room to expand supply and match any price cut, they can scare cheap challengers away from trying to undersell the cosy cartel.

Big business also imprisons information, making it hard for customers and law enforcers to know exactly what goes on inside it, as a by-product of making it harder for competitors to copy what goes on. Using patent, copyright, trademark and other trade secrecy protections, companies can create a legitimate monopoly around their bright ideas. And other people's. One of the most lamented features of technology-based multinationals is the way in which they throw a patent-protected ring even around facts and formulae once held as common knowledge. Inventors of high-yield crop varieties in the 19th century were happy to see nature spreading their seed, whereas those of the 21st tend to slap writs on anyone caught propagating without permission. The Black–Scholes formula for pricing traded options was rushed freely into print by its eponymous authors; subsequent authors of lucrative financial market equations have tended to wait until they've exploited the exclusive knowledge themselves.

At best, keeping high-tech cards close to the corporate chest is wasteful, since many are forced to find independently what one early discoverer could have told the rest. Instead of burning up only one research and development (R&D) budget, each step forward now demands several, as an originator's rivals try to replicate or reverse-engineer the new idea, creating a version close enough to do the job but different enough to sidestep any plagiarism charges. At worst, intellectual property protection costs lives. Premium pricing denies patented substances to millions who need them but whose own (and their governments') budgets cannot stretch that far. South Africa, Brazil and India have recently forced the issue over AIDS-fighting drugs, threatening to buy or supply illegal copies if the patent-holders cannot cut their price in recognition that humanitarian value counts for more than shareholder value. Battles like this have always raged within nations on occasions, but they multiply with globalization, because gaps in the ability to pay are so much greater, and the places where lives are lost are so far away from those where profits are made.

When asked to show why they need to be so big, managers of these monoliths tend to talk about 'economies of scale'. In most product and service lines, double the output can be made at less than double the cost. These scale-related productivity gains begin with the division of labour in production, since workers assigned to narrow tasks become better at performing them than if they tried to master the whole set of tasks, and wasted time switching between them. This division of physical labour is later extended to mental exertions, as administrative and professional tasks are similarly subdivided. Once simplified to economize on labour skill, tasks become more easily transferred from people to machines, so automation becomes a further source of productivity gain as division of labour becomes more detailed. These machines are often costly to install, but their expense is spread more thinly as volume increases, giving another means by which average costs fall as production scale rises.

But the giant size of today's biggest companies cannot be explained by plant-level scale economies, since each product tends still to be produced at several sites. It is hard to explain by scale economies in R&D, which is again typically dispersed across a number of locations. Scale economies in distribution and marketing are similarly belied by the continued use of multiple sales channels, each peddling a variety of products. 'Scope economies', enabling bigger companies to boost efficiency by finding full-time uses for previously under-used resources, have seen their excuse-value drop as most managements renounce diversification and return to core business.

Proving that big firms are more efficient than the small firms they devour or destroy has not been easy, either for the merger and acquisition (M&A) architects who defend their work as 'in the public interest', or for the competition regulators who approve it on that basis. Shareholders have generally been less hard to impress. Although a corporation's share price will typically drop when it launches a takeover bid for another, and the market value of the two put together always seems to start out less than the constituent parts, big buy-ups tend to produce not just a massive bonus for the conquering managers, but also a party-time payoff for both shareholder groups. If greater efficiency is not the reason for post-acquisition profit growth, the next best explanation is good old monopoly power. The absolute size of the merged players allows them to drive a harder bargain with lenders, employed labour and other suppliers, who have diminishing scope to take their trade elsewhere. Their relative size in the market-place similarly lessens the options for customers, unless they achieve a matching concentration of their spending power. Monopoly, once established, tends to spread itself up and across each supply chain, until everyone is either in a group set up to wield power over the market, or part of a defensive alliance aimed at countering that power.

When asked why they need to charge so much for new technology, even when those who most deserve it can thereby least afford it, companies point to the escalating costs of innovation. New drugs cost so much to bring from petri dish to production line, new cars from prototype to showroom, new microchips from drawing board to desktop, that no one would pursue them without the assurance that the first to get a patent can set a monopoly price. Making cheap generic versions available, they warn, would give everyone today's miracle cure but deprive them of tomorrow's even more effective version – because firms would no longer make enough profits to finance the required R&D, and even if they did, the quick drop to competitive pricing would rule out the returns that make it worthwhile to invest.

## Beyond the Brass Plaque

Faced with such heartless business logic, the best way to demonize is to de-personalize. So anti-globalizers' anger is shouted at corporate logos, its brickbats hurled at institutional plate-glass. ChevronTexaco, Celera Genomics and Citigroup give amorphous identity to the burning effigy. And no anti-capitalist demonstration is complete without the ceremonial 'dismantling' of a McDonald's branch. John Deere, Disney, Ford, JP Morgan, Philip Morris and other outgrown eponyms achieve a particular force of impersonality, because the name that adorns their sprawling empires was once a breathing individual. Others achieve the same sense of lost contact with an automated call centre, because within living memory there were human voices at the end of the computer-managed helpline, names you could write to when the database got your numbers in a twist.

The chairs and chief executives who beam down from boardroom tables are rarely worthwhile targets for attack. By the time their mugshot is leering from a poster or their surname woven into a protest song, the chances are they'll have moved on – ditched by shareholders because, from the other side of the boardroom barricades, it seems they weren't ruthless enough in prising out extra profit. Or they're drawn into early retirement to make sure there's time to spend enough of their golden handshake, having sold their shares before the market crashed.

Companies may burn brightly for a time under the command of a household name, but the only ones assured of sustained success are run by people you've never heard of. Charismatic bosses who stay too long invariably fall down or fade out, sometimes destroying the whole edifice when the golden touch is suddenly turned to lead. Studies have shown that, when teams are released from dictatorial command and given power to manage themselves, they impose tougher rules and tighter policing than the slave-driver ever dared to. A watched workplace never boils, because its inmates cooperate in keeping effort just above the bonus-triggering or penalty-avoiding level. Self-managed team temperatures instantly rocket, because every member knows they must work harder to compensate for the unpulled weight surrounding them.

So Ronald McDonald is the public face of the ubiquitous bovine sandwich. It takes a trawl through company histories or Hamburger University reading lists to learn that McDonald's conquest of the worldwide menu was the work of one Ray Kroc. Mickey and Donald do the selling job for Disney Corp. Its turn-of-the-century chief executive, Michael Eisner, for all his billion dollar bonuses, was apparently there just to answer the Mouse's fan-mail and sign the Duck's quarantine certificates. Like many other bosses, they'd discovered that, when wielding such power, it was best to retreat behind a fictitious figurehead. Especially one with a screen-filling smile and lots of fur.

## Too Good to be Traded

Today's typical MNC is owned by institutional shareholders in the rich Organization of Economic Cooperation and Development (OECD) nations, but making significant profits in less prosperous economies. This isn't the only kind of company that can exist, legally or structurally. But it's the only kind with the proven capability to grow large, work across borders, stay consistently in profit, and change its products and services quickly when the old ones cease to pay. The 'Anglo-American' corporation – 'PLC' on one side of the Atlantic and 'Inc.' on the other – has become a uniquely efficient machine for generating the resources for survival, and switching them quickly enough from 'sunset' to 'sunrise' uses.

Firms with other kinds of shareholder (such as cooperatives owned by employees or customers, and limited companies with large family or non-OECD investor holdings), and those with none, tend to be held back by two limitations. There's no one to give them sufficient capital to invest when times are good, or to get through a cash-flow crisis when times are bad. And there's no one to force them to keep doing the most profitable things with that capital, even when it means ditching the people who are currently employed or the customers currently served. Without shareholders to draw on, many expanding companies end up borrowing too much, collapsing under the weight of their debt – sometimes after the investment that they borrowed for starts making profit. Their creditors do, belatedly, exert something of the discipline provided by shareholders, but usually to wind up the struggling company rather than help it get back on the road.

Corporate survival needs owners with a big stick, as well as deep pockets, because we're in a cruel-to-be-kind economy. Just as surgeons must sometimes saw off a leg in order to save the rest of the body, crisis managers take pride in axing half their staff to help the other half keep their jobs. Profit-driven companies sort out their fundamental solvency first, then work out how to ratchet up profit with 'socially conscious' schemes that get as good as they give. This gives them a natural advantage over more genuinely benevolent forms of business, which tend to chase top civic marks at the expense of the bottom line.

This 'creative destruction' produces a better long-term future, whereas shielding current arrangements from shock will only cause inexorable decay. People thrown out of work by one company will find work with another, if they haven't already saved up enough to retire on. Loss of jobs because skills are outmoded is essential if we're to move from carts to cars or slide-rules to computers. Enterprise must periodically go to the wall to clear out antiquated staff and outmoded ways of thinking. Whole economies need occasional booms-and-busts to demolish dysfunctional structures and write off old debts. Responding to shareholder pressure to scrap weak areas and spice up the rest, US companies raised the efficiency of their investments by 40% in the 15 years after 1982. Even in Germany, where shareholders traditionally deferred to hands-on bank creditors, the gain in profit per unit of new capital was more than 12%. Because it gets the most out of society's physical and financial resources, fans of the shareholder business hail it as best for civic and human resources. Workforces may be laid off and closure-towns laid waste in pursuit of efficiency, but at least it gives the extra income base from which governments and charities can try to clean up the mess.

Naked profit-making is never the mega-corporation's only motive. Shareholder-owned companies are famous for their generous pay policies, charitable donations, campaigns to raise workplace quality and reduce environmental impact, and bankrolling of democracy through political party donations. But these concessions to corporate citizenship are of the boomerang variety. Pay is raised if this induces people to stay longer and harder on the job, or keeps the footloose loyal at times when they could easily work elsewhere. The rises are often retracted when a downturn leaves firms wanting to cut production, and staff less equipped to jump ship. Companies drop their usual opposition to public expenditure when it comes to education and training, because the more skills are made available (at public expense), the less it costs to recruit and retain them. In 2002, forced to bring pension fund deficits onto their balance sheets for the first time, UK companies found another way to nationalize their cost base, forcing early retirement on staff whose pensions were linked to their final salary, while simultaneously hiring new recruits who on retirement will get no more than what stock markets have managed to make with their money.

Companies' charitable ventures, while superficially altruistic, slip quickly into 'cause-related marketing'. Donations are chosen according to potential for free good publicity – and sales openings that get much of what's given to flow straight back. When newspapers sponsor literacy classes, software firms give computers to schools, and drug companies distribute chemistry revision aids, the lucky users soon learn to spell 'Ulterior Motive'. You don't have to thirst after knowledge to understand why PepsiCo gives away free salty snacks so close to its soft drink machines. The same commercial logic applies to political campaign contributions, which corporate donors like to portray as subsidizing the process of democracy rather than bankrolling the leaders it elects. To show how scrupulously fair they're being, many companies hand out money to all sides. Their professed aim is simply to save people from paying for the publicity that prises their vote from them. In return, they expect nothing more than a fair fight, and a sympathetic hearing when law-making boots stray onto profit-making territory. Any connection between money passed before the polls and legislation passed after them is a scurrilous confusion of coincidence with what's merely a good cause.

Forms of enterprise without external shareholders – co-operative, privately held, mutual, non-profit, family- or state-owned – invert the 'equity culture' logic. They aim for whatever profit is needed to maximize employee, customer and wider community interests, rather than whatever satisfaction for these stakeholders is needed to maximize profit. However virtuous from a social viewpoint, this makes them vulnerable to economic attack once the game is invaded by seekers after shareholder value. Absence of shareholders leaves them less to draw on when a breakdown in the economy leaves them short of current cash-flow, or a breakthrough in technology gives them lots of extra projects on which to spend. Unless docile lenders or deep-pocketed partners can step in as a substitute for the external capital market, unlisted companies are limited in how far they can upgrade their technology and how quickly they can expand.

In setting out to serve a much broader constituency, not-just-for-profit businesses walk a tightrope between multiple conflicting demands: for higher pay and lower prices, work-place democracy and quick decision-making, the highest labour productivity and the maximum number of jobs. This sometimes disadvantages them in the product market, causing shrinkage and eventual disappearance as quoted companies invade their sales pitch and grab their profit. More usually, shareholder-driven rivals don't make better or cheaper products, but snatch an advantage in the capital market – buying up the troublesome do-gooders with offers that even altruistic stakeholders cannot refuse. This was the way in which McDonald's bit back at Pret A Manger, taking a stake in the group that set out to provide a higher-quality alternative to burgers. And it provided the means for Unilever's midnight feast on Ben & Jerry's, the ice cream that was meant to be eaten while raising two fingers against mega-corporate malnutrition.

## Multinational History: First Fords to Fast Food

Multinational companies' main aim is to survive and grow. For this they need to consume a continuous flow of raw materials, topped up with the occasional battle-slain competitor. Like most on such a demanding diet, they prefer to take these meals at home, or in familiar restaurants close to it. So those early colonial incursions into the lands of sorghum, soya and spice gave a wholly misleading picture of their long-term plan. But first impressions count; and in the MNCs' case, they were generally the worst impressions – raiding the wilderness for cheap raw materials, fuel and labour, whose annexation cut their costs and raised their profit rates; then returning home with the threat of low-cost recruits in rival locations, to force down pay and raise their profits higher still.

A century ago, large tracts of the planet already feared being Shelled. The big oil firms' killer drillers would come and buy up land whose oily texture had prevented its cultivation. Or they simply fenced off rocky outcrops which no one had thought to put to private use. By spending vastly more than any local could afford, building machines they knew nothing of to cart off a juice they could do nothing with, the hydrocarbon colonists gave every impression of a rescue mission. By the time their hosts saw what was in the pipeline, it was too late to turn back the extractive tide. A petroleum-based economy was already taking shape, with the rubber, chemical, car, truck, transport, logistics, steel, rare metal and finance multinationals ready to follow down the internally combusted trail.

From this grim start grew the image of the big multinational as planetary predator, swooping wherever there were cheap minerals to extract or cheap shoulders to press to the grindstone. Not content with driving down costs, the MNC extended its grab to free resources: seeds and cells to be patented, soils and skies to be polluted. The rise of the corporate conqueror seemed surpassed in rapacity only by the rises enjoyed by the fat cats who planned its advance.

But these extra-territorial excesses were an early aberration, and one the MNCs have since been trying hard to live down. Few ever wanted to leave their polluting or brutalising bootprints in foreign mud, because few ever liked stepping out beyond the tree-lined streets of home. MNCs were always upset at having to go abroad for their vital inputs, and worked remorselessly to bring their 'sources' closer to home. Burying oil, rubber, aluminium and other industrially essential metals and minerals in impenetrable jungles, deep oceans or parched desert sands was one of nature's crueller tricks. It was one that the first commercial exploiters quickly tried to correct, by developing locally procurable substitutes, or bringing home a stock which could thereafter be relentlessly recycled, so as to minimize the need for replenishment from source.

Having paid such a price to bring them together, making profligate use of scattered natural resources was not what multinational managers wished to be famous for. Making economical use of concentrated human resources was not only easier and cheaper; it also moved the core concern from volatile primary commodities to manufactured products in much more stable demand.

Big businesses from big nations like the US go abroad primarily to strengthen their hold on the domestic market. Like rival generals who agree to fight on neutral ground to avoid destroying the riches they're battling for, rival companies go abroad for extra profits, people and product ideas to bring to the domestic battle. When the going gets rough, the wounded warriors retreat back home, even if that's where the bloodbath is at its fiercest. When such US globetrotters as Chrysler, Safeway, AT&T and Woolworths over-reached themselves in the 1970s, they sold off foreign assets. Taken over and turned round, these became same-name rivals that later stood in the way when the fair-weather parents attempted to re-enter. Showing that this century's industry captains learn no more from their history than infantry commanders of the last, the same instinctive pull-back began when America's latest boom faltered in 2000–01. Merrill Lynch ran down its long-running efforts to enter Japan; Motorola and Gateway gave up most of their European efforts; energy and water utilities sold their British holdings to local rivals for usually a fraction of what they'd paid at privatization.

The motivation is different for big businesses from small nations, or those that perceive themselves as small. Sony and Toyota from Japan, Hong Kong's HSBC and South African Breweries are among MNCs whose restricted home patches have provoked a much more outward-looking approach. They move the focus of their operations, if not their headquarters, to wherever in the world they see the brightest long-term future. They step up the switching of business abroad when their domestic sales turn sour.

By early in the 20th century, multinationals – and their critics – were already less worried about the source of material and fossil energy inputs, and more concerned with what became of them once brought back home. A mass consumer market was developing in the US and other high-income economies. As manufacturers strengthened their power over primary input producers, and then retailers flexed their muscles against supplying manufacturers, the main source of profit moved 'downstream' from the raw end to the more refined middle of the supply chain. Employers' enthusiasm, and workers' fears, shifted towards Fordism. Named after the car-builder Henry, and often linked to Frederick Taylor's 'scientific management', Fordism set out to rewrite the industrial rules. Work was no longer a source of fulfilment to which everyone had a right, but a duty which could never be inherently enjoyable. The employer's task was to make work hours as short and productive as possible, to leave more time for having fun outside the factory gate. At the same time, profit was no longer maximized by setting high prices and selling only to an élite. With production lines delivering economies of scale, margins went higher if prices were dropped and production stepped up to serve a mass market. Mass production summoned mass consumption, because a more productive workforce had extra money, and the time in which to spend it. Well-paid workers could accept empty hours in the workplace in exchange for stacked shelves in the market-place. Labour unions' efforts to win collective benefits, as producers, were forced on the defensive as members looked instead for individual victory as consumers. On the production line, and in the typing pool, employees' fight for extra rights was undermined as task simplification made them eminently interchangeable and replaceable.

Ford's original aim had been to expand his own market, raising workers' pay in line with their inexorably growing productivity, so that they could always buy the extra goods they made to sell. Although real wages rose under the Fordian bargain, even without a push from strong unions, productivity kept pace, so profitability kept growing. The vision would eventually have stumbled on the industrial economy's inability to make demand keep pace with supply, a problem explored further in chapter four. But war and recession in the first half of the 20th century had already alerted other entrepreneurs to the modern economy's proneness to phases of slower growth and periodic retrenchment. No longer confident enough to plan production on the basis of a continuously growing economy, business turned to ways in which it could continue to grow in conditions of stagnation. Emphasis shifted from raising the value of output to reducing the cost of input, and producing the same with less rather than more with the same. So by the 20th century's end, big business's colour and shape had changed again, from blue oval to golden arch.

Organizational initiative passed from Fordism to McDonaldization, as structural changes in producer power tipped the advantage from oil barons to burger kings. Industrial legend has it that Henry Ford dreamed up the assembly line after seeing the process run in reverse at an automated abattoir where dead animals were dismembered in a scientifically determined sequence of steps. As the carcase was torn apart, so the car was to be welded and bolted together. With Ronald McDonald at the helm, de-skilling spreads from car assemblers to butchers, bakers and burger bun makers, as basic nourishment slips from biological to chemical and slides down the assembly line.

Ancient arts of food preparation and presentation surrender to marketing science, in the guise of alchemical grills transmuting bonemeal back to meat, and ovens programmed never to burn the fries. Compared with the Fordist workforce, McDonald's employees are remarkably highly qualified. But that's mostly because of training they received before succumbing to their contracts. Casual jobs at fast food outlets seem to be the income support for at least a third of the world's college students, as well as a standard first job for those who graduate in philosophy or film and media studies.

On the other side of the counter, consumption is equally robbed of its craft, as acts of eating and drinking revert to unthinking snack-time sub-routines. Ingredients so highly processed that they melt in the mouth and don't re-solidify until the end of the intestine are washed down by beverage so soft there'd be more calories in munching through the straw. After the infamous incident when a spilt McDonald's coffee earned six-figure personal injury compensation, there are even printed instructions (far more nutritious than the wrapper they're written on) for how to hold the cup, swallow the cola, slice the bread, and slurp the last milkshake residue without drowning out the piped music.

Despite their outward similarity, and frequent proximity, the hamburger joint is a long lunch hour away from the car assembly plant. Whereas Ford sells a product entirely defined by bodywork, chassis and engine, McDonald's sells a service that entirely transcends its physical ingredients. Location and lighting are all part of the paid-for experience, to the point where the mechanically recovered meat-paste on the table is almost incidental. So while competitors can break Ford's product down for ease of imitation, McDonald's remains a whole that defies disaggregation to constituent parts.

Factory farming ensures that meat, bread and compulsory grease comprise only a fraction of the burger's price. Cheap energy policies pare frying, baking and reheating costs down to an equally derisory sum. So more than ever, labour is McDonaldism's main expense. Because its delivery mechanism must go wherever the rumbling stomachs are, the fast food empire expands by marching across borders. Franchising speeds this wildfire spread by making each outlet take on the costs of investment, and associated risks. And this branching out sets the scene for a borderless 'benchmarking' exercise, ratcheting pay down and performance up to the keenest possible levels. Within each outlet, the workforce has to settle for minimum pay, to fend off replacements who will stoke the eternal fires for less.

## Making Markets

Governments which had tried to build their economies on private competition ended the last century confronting a basic dilemma. The free play of market forces had conjured up a set of corporations so powerful that they seemed to tie up the 'invisible hand' that created them. Big businesses were such concentrated players in the product markets, they could name a price to scattered, disorganized consumers. So focused was their hiring power in labour markets, they could dictate a wage to similarly dispersed, de-unionized workers. Competition among many uncoordinated producers and consumers, which economists showed was vital to the market economy reaching an efficient equilibrium, had been squeezed out of the market economy by its enterprises' own success.

It is unfair, however, to accuse giant firms of trading roughshod over the free-market playing field. This assumes that a world of fair competition and full information once existed. History questions this, and commercial logic effectively rules it out. Completely free markets would be wildly unstable, with customers leaping en masse from one product to another whenever their relative prices changed, and similar stampedes on the supply side if one activity ever became marginally more profitable than another. Innovation would be impossible, because any new product or process would be instantly copied, denying its originator any payback on their effort. Completely free markets would get stuck on a high unemployment and low growth path, as businesses failed to invest as much as households wanted to save (through mechanisms explored more fully in chapter four). Ground zero would eventually be reached, as perpetual competing-away of profits made managements realize that there was no point in staying in business. In a perfectly competitive world, shareholders could get the same return for less risk by keeping their cash in a sock, if only it were worth someone's while to make the sock in the first place.

To keep going, markets need support from social structures and behavioural conventions that have to take shape outside the market. Left entirely to themselves, people can't work up the shared knowledge or build up the trust which make it safe to specialize, and so link their survival to market exchange. The fate of a completely free market system was chillingly illustrated by Russia, in the first five years after the fall of communism. Mafias took over from a police and judiciary that no one trusted, barter replaced trade with a currency that no longer existed, and the private sector failed to regenerate any of the jobs lost by a disappearing state, because the only profitable companies were those which held on to a legally ring-fenced monopoly. Western advisers who had urged the wholesale abandonment of Soviet central planning began bemoaning the sudden loss of its 'institutional quality', and turned to China – whose private enterprise took shape under careful Communist Party guidance – as the favoured example of how transition to free markets should really be done.

Far from demolishing market exchange, big firms often rebuild it in the only way possible. Like any contest which gets called off whenever the rain falls, the obvious solution is to bring it inside. Once every production plant belongs to one firm, competition can be unleashed when it's useful and turned off when it begins to get destabilising. People and subdivisions can vie with each other when the competitive edge needs sharpening, but work together when there's a change of product or procedure to agree. Once every employee is under one boss, information can be pooled, achieving the other condition required for free markets to function. If everyone knows what others intend to do, and what actions they can draw on to do it, they can quickly arrive at a consistent set of strategies. Rescued from a commercial battlefield where they must hide or misrepresent their aims and abilities, people can make sense of their surroundings and plan their next move. Coordination by the corporation avoids the competitive second-guessing and frustrated expectation that otherwise leaves consumers tripping up on misinformed decision, and producers paralysed by reluctance to reach any decision at all.

Economists don't like admitting that markets work better inside big firms than outside. They rationalize the multinational's rise as due to 'organization-level' scale economies. Bringing lots of operations under one corporate roof is meant to allow investment to be raised more cheaply, coordination to be improved so that duplication of effort is avoided, new ideas to be shared out so that they find a money-making application more quickly, and resultant new products to be designed, built and marketed simultaneously so that they arrive on the market too quickly for copycat products to get there first. But these are all the result of fusing market discipline with the hierarchy's ability to decree and dictate. Efficient allocation of *decision* resources, not of human or material resources, is the reason why big is financially beautiful. Large companies survive and thrive because they are market systems floating in an ocean of corporate planning – precisely the reverse of the commercial universe that commerce's cosmologists would like us to observe.

## The Planetary Planners

Making the market an obedient servant, without succumbing to its mastery, has been a dream of governments throughout the industrial age. But it has been left to the modern multinational company to achieve it. Only inside today's big business has a workable mix of galvanising competition and stabilising coordination finally been found. With it, today's large firm can make its environment sufficiently predictable to make major, long-term investments with a certainty of payback. It can ensure that innovation is worthwhile, by setting prices that cover the R&D cost. It can make sure that every product finds a market, by lining up buyers in advance.

More remarkably, it can do all this far more quickly and reliably than most other types of organization, especially the organization – government – traditionally charged with taking major economic decisions, and entrusted with the powers to carry them out. A corporation's strategy committee is quite capable of closing a division, authorizing a new product development, changing the corporate logo and sacking the chief executive in the time it takes the equivalent cabinet sub-committee to schedule its next meeting. The new imperial company can shed its Latin American operations, reconfigure Asia, open a strategic outpost in Africa and double its European capacity, while the old imperial country is still trying to sort out the mess left by the arbitrary borders that it ran around restive colonies a century ago.

In the early 1990s, Russia's new prime minister, Yegor Gaidar, launched a 100-day plan to rescue his failing multinational enterprise. Communism's internal organization was to be shaken up by a radical decentralization. Departments previously string-pulled by the centre would now be free to set their own rules, putting responsiveness to local conditions before conformity with any global grand plan. Prices once laid down by head office accountants, leaving clients buried in butter and bereft of bread, were to be set according to current market conditions, supply and demand taking over from high command.

Fast-changing tastes and fast-transmitting information technologies were recognized as having sunk the old top-down model, ushering in a necessarily more decentralized age in which multiple, diffused political centres worked through new, undirected networks to bring order to a world no longer ruled by just two mutually disciplining superstates. Russia's external engagements were to be radically scaled down. Afghanistan was already abandoned, and the former Eastern Bloc satellites spun out as independent outsources were now to go their own way as clients or competitors. An old Cold War understanding with the long-time arch-rival had been renounced, with more incandescence than a thousand burned American flags.

At around the same time, Coca-Cola's new chairman, Robert Goizueta, was launching a similar 100-day rescue plan for a similarly troubled worldwide enterprise. Coke's internal organization was to be shaken up by a radical centralization. Departments which previously set their own rules would now be ring-pulled from the centre, putting conformity with a global grand plan before responsiveness to local conditions. Prices once set according to current market conditions, palatable for the buyer but distinctly unprofitable for the supplier, were now to be laid down by head office accountants.

Fast-changing tastes and fast-transmitting technologies were recognized as having sunk the old bottom-up model, ushering in a necessarily more centralized age in which single, central islands of power vied through new, directional networks to bring order to a world now ruled by just two mutually disciplining soda makers. The company's external engagements were to be radically built up. Asia was already on stream, and the former Eastern Bloc satellites spun out as independent outsources were now to be brought on board as colleagues or cooperators. An old Cola War understanding with the long-time arch-rival had been rejected, with more discordance than a thousand crushed Pepsi cans.

Ten years on, Goizueta's company, still a largely US-based enterprise until the early 1980s, sells to more corners of the world than ever before. Gaidar's country, still a global presence as that decade dawned, now cannot control even the stroppier secessionists within its own diminished boundaries. A shadowy formula for sugary water has conquered the globe, while the Enlightenment formula for socialist welfare has shrunk to one island in the Caribbean. While the communists fell down through a failed adoption of the free market, the Coca-colonists stepped up through a successful reinvention of central planning. Companies like Coke can succeed, where governments have failed, in using controlled market forces to keep grand plans on the rails, because of their more focused mission, clearer constituency, and better decision-making capability.

But to recreate markets in the only place that they really work – inside the corporation – companies must not only grow big. They must also go multinational – precisely the tactic that they had been retreating from, ever since the first ill-starred colonial incursions. Companies had neither ready means nor obvious incentive to force their way back into foreign markets, until non-market forces pushed them over the edge. The source of those forces is the opposite of what MNC critics conventionally allege.

## Who Let the Dogs Out?

Corporate-driven globalization is a political creation. Most companies could not go abroad until governments created the conditions that allowed them to do so. Most did not choose to go abroad until governments forced them. Far from inflicting their expanded presence on an unsuspecting world, big businesses grew to fit a role in which others had cast them.

Industrial country governments that had carefully dismantled their trading empires did not expect their companies to stride straight out and recreate them. Their motive was precisely the opposite. By consistently outgrowing the economies around them, leading companies have built up monopoly buying and selling power over product and labour markets, even in the world's largest economies. Taking away external trade barriers was intended to dilute that monopoly power, by throwing the big fish into a very much larger, global pool.

The defence proved a disaster. Big fish thrown into larger pools don't thrash around in terror for long. For the new MNCs, the global plunge gave an especially large boost to power because of the chance it opened up to internalize market forces. The swing in economic influence from state to enterprise was especially great because governments were giving up the planning powers that private enterprise was taking on. Impressed by the ability of private enterprise to get more done with fewer resources, governments were rushing to privatize their own industrial assets, and subject state agencies to private management. And government had to make do with fewer resources since private enterprise was using its power to withhold them. But this message got through too late to stop the state from abandoning external trade defences that were also the domestic market's shield.

The state's desire to defend free markets against the enemy within is understandable. The market is very much a political own creation. Economists have tended to argue that free trade is a natural state, and government an intrusion that at most times disrupts it. But if state intervention kills private enterprise, it is not before bringing it to life. Public authority is needed to enshrine the property rights traded in markets; to establish the currency in which they trade; to defend that trade against coalitions' attempts to manipulate or corner the market; to stabilize the flow of trade against the market's characteristic fluctuations; and to provide institutional supports like road networks, policing, lighthouses and law-courts, which market traders need but have no incentive to privately produce.

Governments had acted to contain commercial forces that were endangering competitive economies at national level. Instead, they had let the global capitalists out of the bag. Ford would not have gone abroad if state-imposed trade protection hadn't made it cheaper to export whole factories than single cars, and McDonald's could not have done so without state-agreed means for repatriating income and respecting franchisees' rights. Grasping why governments took this false step means puncturing some myths surrounding anti-globalizers' second major target: traditional theories of the benefits of international trade.

CHAPTER TWO

# Trading Insults

*The all-round gains from free trade are at the centre of arguments for corporate-led globalization. But markets don't work in the way they're meant to in conventional trade theory, and no nation has got rich through specialization in the way that the theory suggests. Big corporations, far from subverting competitive markets, are the only force that can make them work. They have used this ability to build a central planning system far more effective than any achieved by socialism. Globalization emerges as governments' only hope of keeping the economy safe for giant enterprises, as well as keeping the people safe from them.*

The technological simplification of life, leaving more time to dwell on its logical complication, propels increasing numbers of people into a search for higher meaning. This tends to end halfway up a hotel-owned Himalayan mountainside, where a German-sponsored guru in a Saudi-fuelled motor-home sips American soft drinks and surfs the remnants of the dot-com boom on a Taiwanese-built notebook. Seeking solace in authentic South American jungle sound, disillusioned trekkers must shield their ears against the thud of Chinese-made trainers, executing European dance manoeuvres to celebrate the union of Japanese stereo and Anglo-Indian pop. The cosmopolitan conspiracy pursues them home, where even the remotest village pub can summon spirits from five continents. Well, unless they drop in on my local. There, for some reason, even Newcastle Brown Ale is viewed as a mystery potion from a distant, alien world.

Nothing we do is truly local any more. All gets shaped by – and rebounds on – the thoughts and actions of others, including people we've never met, in places we never go. Even local radio hums with World music, interspersed with words on loan from other languages and ads for products that no longer come from just down the road. Even local heroes need a Hawaiian tan and a Hollywood smile, and a private beach in some exotic retreat on which to be caught by the long-lens camera. Local buses have ceased to exist, but there's probably a flyover for some long-distance motorway not far from you. Just after I wrote that snide remark about my local's parochialism, it was bought by a nationwide pub chain with the help of a Japanese bank. Now it's just switched from real ale brewed in the UK by Danes to palpably fake ale with a Czech name but full of North American lack of flavour.

What we said and did used to echo not much further than the garden gate, or the other side of the table. Now everything bristles with interconnectedness. You can't even bet on the local dogs without worrying about the impact of new Middle-Eastern crossbreeds, or retire without a pension whose payout is endangered by the downturn in North Asian equities and Latin American bonds.

New Yorkers felt comfortable in their wealth and proud of their way of life in the Fall of 2001. This didn't shield them from airborne visitors who took a dimmer view of their high-rise living, believing their own material and cultural standards to have been dragged down by America's ascent. In turn, the hijackers who flattened the World Trade Center couldn't escape a wounded Pentagon's vengeance, even after fleeing into Afghan caves. In the Cold War era, countries unwilling to be chips off a superpower bloc could declare themselves non-aligned, or spin the giants into a bidding war, and so do their own thing while missile crises raged above their heads. In the new century's 'war against terror', the choice is one global alliance or another. Staying at home is not an option, and little local difficulties – a mistimed handshake in Jerusalem, a mistranslation in Seoul – can send out planet-circling shockwaves.

## The Fatal Connection

*Social* interconnection makes it impossible to enjoy being rich without acknowledging an unpaid debt to the poor. In every age, a handful of 'haves' have wined and dined while an unseen mass of 'have nots' toiled away to stave off starvation. But ruling-class consciences were always kept clean, and their treasure stores kept safe, by the huge distance between the bulging plate and the begging bowl. Far from setting up inevitable confrontation, the sheer disparity between élite affluence and mass squalor gave a form of protection. Rich and poor achieved a peaceful co-existence by literally living in different worlds. It is the closure of that gap that has opened the rifts. Now, the over-fed executive might actually have a way of wrapping up the excess dessert and sending it to a shanty town. Subsidies designed to prop up prices for pinstriped farmers in the Northern hemisphere really could be better spent supporting basic incomes in the South. The price cut enjoyed by clothes shoppers in Manhattan really might leave a paper trail straight to the wage cut suffered by textile workers in Bangalore.

*Economic* interconnection stops us being the sole authors of our fate. Once we enter the division of labour, we depend on others for the materials that we work with and the markets we sell to. One region's rising tide can lift the boats in many neighbouring harbours. Equally, an America-sized sneeze is no longer needed for the rest of the world to catch cold. The collapse of a single over-borrowed country (Argentina, Indonesia), or even company (Long-Term Capital Management, Enron), can shake it, as the losses of its creditors and customers spread knock-on 'contagion' across financial markets. Lightning strikes in California, or wildcat strikes in Korea, can pull the plug on just-in-time production across Europe. Homegrown labour unrest is no longer needed to bring activity to a halt.

*Ecological* interconnection rules out revelling in command over natural resources, by pointing to others' – and our own children's – captivity to them. Fuel burnt in inter-city traffic queues is snatched from under desert-hardened camel hoofs. Its exhaust clouds spiral into mountain-village air. One nation's bad breath is another's acid rain or adverse climate change. But emission-cutting initiatives may hurt only their authors, unless everyone adopts them. So Europe and Japan are now hesitating over commitment to the Kyoto greenhouse gas curbs after their unabashed rejection by the US. Even establishing that global warming and ozone depletion are actually occurring, and necessarily harmful, is a border-hopping effort, relying on measurements and modelling from all over the sweltering globe.

*Cultural* interconnection stops us escaping these awkward questions through a retreat into shared backgrounds of well-worn words and handed-down sound. The books we read are now inspired by incidents on Hungarian-built buses and printed on presses in Hong Kong, even if they flow from the quill pen of an author who's never ventured outside Houston. The films we see are given American scripts, romantic Italian leads and scene-shifting African rhythms, whether the storyline is borrowed from Norse tragedy or Aboriginal myth. The songs we hear are most likely to be recorded in America, put on disk in Bulgaria, replayed on Chinese-made hi-fi and overheard in a Danish inn, notwithstanding that the singer is Estonian with a backing band of Finns and Greeks.

Interconnection feeds on itself. When we depend on others – and they on us – for finding things out and getting things done, one missed or mistaken connection can blow everything off course. 'Chaos theory' has alerted us to the fortuitous but fatal progression from the flutter of a butterfly's wings in China to the breaking of a tropical storm in Chelmsford. Bill Clinton's presidency set up a similar link between the flutter of an intern's eyelashes in the White House and the dropping of bombs on Iraq. The same inescapable connections now recur whenever people reach us from a distant departure lounge, or goods through a global supply chain, or information via satellites that don't see lines on a map.

The interwoven world is popularly characterized as a 'risk society'. We are shown to have dispelled the uncertainties of ignorance (resolving thunderstorms into atmospheric fluctuation, illness into chemical reaction and the wrath of the gods into psychological disturbance), only to have created new uncertainties through the pooling of effort by which our earlier helplessness was erased. Things we once knew for certain, because they were fully under our control, are now thrown open to doubt, because someone else somewhere else must move the right levers or pull the right strings.

Once set loose, connected society's risk can't be contained by just severing links. Defusing it means passing it on to others, or spreading it around. So we combat the stresses of interconnectedness by establishing more interconnectedness. Investment risk is tamed by fanning out the funds into a global portfolio, relying on a boom in one country or region to offset sudden downturn in another. The risk of stock-outs and power cuts is solved by drawing our product and energy supplies from sources that are far apart, geographically and diplomatically. Our escape from the risk of social rejection is to have friends in highly different, distant places. Once we take a single step down the tangled path towards interconnectedness, trouble only forces us to take further steps. You can fight bad connections with new connections; you can't ever reel them back in.

## Getting Netted

Interconnectedness finds high-tech expression in the concept of a 'network society'. Top-down connections, forced on people by arbitrary inequalities of status and power, are supposed to have given way to voluntary, two-way exchange between people who share a commercial or cultural interest. Previously dictatorial governments are supposed to become more democratic, hierarchical organizations to flatten out, and father-dominated families to give way to those in which 'new men' share the domestic tasks, under this new multi-centred arrangement. The network society takes on high-tech form in the inexorable rise of electronic networks. Whereas old views of the world as a mechanism gave people a deterministic picture of where it was going – and made them keep taking wrong turnings – new views of the world as a network alert us to its openness, and to the liberating discovery that neither over-mighty gods nor over-confident planners can push us in directions that we don't want to go.

Admirers of these networks believe that they finally deliver one of modern society's holiest grails: distinction of roles without gradation of rank. On the Internet, and its constituent websites, newsgroups and chat-rooms, people can explore and exploit their differences in ability and interest without imposing judgement on their worth. Members are valued for what they know, not who they are. So a managing director can end up taking advice from a maintenance fitter, or a bank clerk giving orders to the finance minister, if this is the flow that ideas and information dictate. The image of equal-status interchange is reinforced by the Internet as a 'network of networks' in which all participants' personal details can be hidden behind a screen name. It is further fuelled by TV terminology in which 'network' denotes the channels that are open to all, not those piped to a privileged few who can afford to pay-per-view.

This picture is, in practice, as far from the truth as most Internet chat and docu-drama dialogue. Only one type of net is truly non-discriminating, and that's the type of fishing net that's reduced the North Sea to half a shoal of sardines. All other networks fall prey to 'networking', the process by which privileged members fight their way to the centre and join the other dots into an intricate chain of command. Network centrality is seized by those who can turn the more distant members from names into 'nodes', providing and processing their information and obeying their instructions. Such corporate superstars as Jack Welch at General Electric and Percy Barnevik at ABB were supposed to have run their organizations as networks, breaking down old communication barriers due to different locations and job titles, and giving everyone an input to decisions that concerned them. But few who worked in these companies, or spent any time with those who ran them, came away with any doubt as to who sat astride the network and whose strategy (commercial and personal) it served.

The tendency of some members to be more central than others is even more obvious with political and criminal networks. When taking arms against the al-Qaeda terrorist network in Fall 2001, the US not only tracked down the network's main headquarters to Afghanistan's Tora Bora caves, but found a single individual, Osama bin Laden, to be the string-puller at the centre of the 'axis of evil'. In *Metropolis*, Thea von Harbou's classic dystopian vision of a post-industrial city, enslavement to the iron logic of efficient people-management is made more chilling when one organizing wizard is revealed to be in overall charge. Those who can get the network revolving around them, and set up as gatekeepers deciding who gets into it, can wield more power in the seemingly 'flat' organization or political structure than the strong arms of their 'hierarchical' predecessors ever enjoyed.

Inequality of wealth and influence does not diminish when old power structures give way to new networks. Disparity may even grow more sustainable, by being made less visible. Networks blur the gap between rich and poor, because people and nations can swap places in the prosperity league more easily, and all can now expect a basic level of provision. Even the street-dwellers have round-the-clock radio, and colour printing on their cardboard box. Even the meanest shacks sprout aerials and satellite dishes, so all can tune to Hollywood gossip and CNN. The products and practices that spread most readily from the West to the rest are not the fast cars and flat screens that symbolize its economic edge. They are the hamburgers and fizzy drinks that show modern life at its most elemental. We have a McDonald's to drive into today, while still awaiting delivery of the Mercedes to drive into it. On such reassurance, of present deprivation being down-payment on future reward, must the most distant players tolerate their entanglement in the growing global net.

Anti-globalizers' demonstrations can't escape this slippage from equal treatment, since some members of their own networks are demonstrably more equal than others. Those who practise what they preach, keeping their loyalties local, risk rapid eclipse by those willing to dabble in the lifestyles they condemn. Planes, trains and coaches speed these professional protesters between global summit sites, despite long-distance travel's social and natural destructiveness being one of their main complaints. Their defence of community against telecommunicated tyranny is spread through telephone wires, or downloaded from awkward corners of the World Wide Web. Their end-century demonstrations made concrete, and often conveyed with concrete blocks, the violence they accused global commerce of inflicting in the abstract. Until, in September 2001, a greater violence stunned the street protesters to silence, the stop-the-world bandwagon itself finding the way blocked by two screen-freezing dust-clouds in the North American sky.

## Commercial Ties

Terrorists chose to strike the World Trade Center because its towers were tall, twinned, and lacked the usual structural supports. But their target was deliberately chosen to resonate in other ways. For it is on the proposed benefits of world trade that the case for globalization is traditionally centred. Drop protective barriers, the free-traders argue, and other global differences will dissolve alongside them. Take off barriers to capital mobility, and the rich world will find profit in reconstructing and developing the poor. Nations will discover a mutual interest in launching container ships in place of gunboats. Geo-political competition that invariably ends in war will give way to economic cooperation that delivers and depends on durable peace.

In the free-traders' framework, a global division of labour harnesses everyone's efforts to the same production machine, giving more to consume than if all kept toiling within national borders. Global product markets help them swap the proceeds, bringing chocolate to the poles and ice cream to the tropics. A global pooling of ideas helps everyone jump to the best practice, instead of working within touristic but time-wasting local traditions. Global financial markets collect the cash that's idle or under-used in one place, and put it to good work in another. A global labour market will ensure that the engineer in Dakar or Dar-es-Salaam who performs as well as her Detroit equivalent will be equally richly rewarded – through African and Asian pay being levelled up, not through the first-world wage packet becoming another American icon razed to the ground.

'Make Money Not War' is not a choice with a very long history. Until recently, trading of blows and of cargoes was engaged in simultaneously, or in frequent alternation. From Conquistadors storming South American shores to Black Ships circling Tokyo Bay, armed forces paved the way for market forces, the merchant ships arriving only when the warships had done their work.

Of many good theories that never quite made it to the Millennium, the McDonald's Law of International Deterrence is especially sorely missed. For years the pioneering fast food chain could claim to be the key to world peace, because no two countries which hosted a branch had ever gone to war against each other. Then Slobodan Milosevic plunged Serbia into a war with an Anglo-American-led alliance, and Nato bombs started sprinkling novel seasoning on Belgrade's Big Macs. Theoreticians still seeking a link between conflict prevention and convenience food had to switch allegiance to the slower-expanding Wendy's chain, or turn to Ikea furniture stores as peace-keeping's more prosaic highest common factor.

Despite this setback, most pro-globalizers view the Balkan conflict as the exception that proves free trade's dispute-resolving rule. The more nations exchange useful goods and services with one another, the less prone they seem to swapping diplomatic bullets and incendiary bombs. The last time the world avoided total war for so long – 40 years up to 1914 – was also the last time it approached present levels of economic integration. World trade flows (the sum of merchandise imports and exports) as a fraction of world output, at around 20% in 2000, were already well above the proportion in 1913, and probably ten times the level to which they had sunk by 1920.

Trade's advance may have been as much the consequence as the cause of war's absence, but their mutual reinforcement seems undeniable. When, after the 1914–18 bloodbath, nations tried to rebuild behind protective barriers, their progress resembled their propaganda about the enemy's tanks: one forward gear and six reverse. Trade restrictions to defend unrealistic pre-war exchange rates between national currencies led directly to the 1929 Wall Street Crash and 1930s Great Depression. Recovery was then delayed because nations could not work together to rekindle activity, instead trying to revive their own economies by raising trade barriers and devaluing their currencies. This exported their unemployment problem, along with a temporary surge of foreign goods sales, but the world fell back to square one (only with more inflation of prices) once everyone had matched the exchange rate reduction. Nations were ultimately driven to seize by force the resources they could no longer obtain by trade, so hastening a Second World War only two decades after the First.

Pre-1914 levels of trade, in relation to national income, had largely been regained by 1980, and the past 20 years have seen most regions reach unprecedented levels of international integration. Modern economies' degree of openness becomes even greater when allowance is made for the shift of production towards services, whose limited exportability makes less of the total available for trade. The proportion of US tradeable goods that actually flow across borders, around 13% in 1913, was already closer to 30% by 1990. And the US is one of the world's more 'closed' countries, especially when primary commodities (crude oil in, raw foodstuffs out) are left out of its accounts. Small, open economies like Singapore, Puerto Rico and Hungary, which colonialism largely kept out of existence before 1914, regularly trade well over half of what they produce.

Global commercial integration seems even more exceptionally high when multinational companies are brought into the picture. MNCs produce abroad as an alternative to sending exports from their home country. Adding these to recorded trade flows boosts the effective level of bigger nations' imports and exports, as well as often reducing or reversing the measured imbalance between the two. As well as accounting for at least a third of recorded international trade transactions, MNCs make significant unrecorded exchanges through their internal transactions. Spinning-out of intermediate production and assembly to foreign outsources, and cost-price transfers between foreign subsidiaries, means much that is external to the national economy now shows up as internal to its firms. When such non-trade transactions between them are considered, even superficially self-sufficient large, rich countries take on a more open air.

## Tricks of the Trade

Growing trade is linked to rising prosperity, economists argue, because integration lets everyone focus on what they do best, and reap the full reward. So as long as nations differ in their endowment of natural and human resources, or the efficiency with which they use them for different purposes, there is scope for specialization that involves all and benefits all. An international 'division of labour' can develop which lets people exploit their strengths, and push up their incomes, as powerfully as the age-old division of labour (butcher, baker, website designer) within nations. For Europe to import Asian rubber and steel to build cars, or South America to assemble computers with microchips sourced from the North, is a natural step forward from crop growers selling their grain to a specialist miller, or herders turning to a cartwright for a better set of wheels.

International trade began with a handful of countries exporting manufactured goods and the rest selling raw commodities that were used in their production. Sometimes the commodity export was voluntary, because nations with a farming or mining base wanted factory goods that only the industrial world could offer. Often it was forced on countries which didn't want any industrial products, but whose natural wealth the rich world still wanted to feed its own material needs. This pattern of trade could easily be explained by saying that non-industrial nations had 'comparative advantage' in natural commodities. The rest, without such rich seams and soils to dig, had to forge their own advantage further down the 'supply chain'. But in creating new demand for the output of farms and mines, and putting an international market value on them, this division of labour between an industrial North and pre-industrial South was argued to bring prosperity advantages all round.

Today's trade pattern is more diverse, with two-way exchange of manufactured goods within the North, an increasing export of transplanted manufactures from South to North, and corresponding export of tradeable services from North to South. Free trade's defenders present this as proof that specialization by comparative advantage not only brings immediate gains, but also amplifies the benefits over time. Rather than having to keep selling commodities and components for others to refine or fashion into finished goods, pre-industrial exporters can 'trade up' to products with higher added value. Among the manufactures that the industrial world competes to sell are the machines and technical knowledge that its factories are run on. By acquiring these technologies, and putting its lower-waged labour to work on them, the developing world can push its own comparative advantage up the value-adding scale.

A long-running criticism of free trade theory, echoed in today's anti-globalization protests, is that the goods only flow because the people aren't allowed to. Because workers in commodity-exporting nations generally earn only a fraction of the wage paid in industrial exporters, their best way up in the world would be to go to where the high-income jobs are. Free-market *economics* demands unlimited mobility of labour within each nation, so that its entry to world trade can be followed by wholesale closure of its 'uncompetitive' industries and redeployment of people into those where advantage lies. Yet free-market *politics* demands strict limits on mobility of labour between nations. High-fliers who can do a job that the local workforce can't, and low-stoopers who can do one that they refuse to, are grudgingly admitted – though often as 'guests' who lack the rights of permanent residents. Anyone else who pursues the profit motive across borders is classed as an 'economic migrant', their would-be host's generosity not extending past a one-way ticket on the next plane home.

Far from stumbling over this disparity, free-traders regard it as a triumph. Their theory shows that when goods and services cross borders, people don't have to. Staying at home and specializing in areas of comparative advantage will cause their wages to converge with those of counterparts abroad. Convergence will occur just as surely and swiftly as if they'd gone abroad for more pay, because the mechanism is the same. Labour moves from where it is plentiful to where it is relatively scarce, closing the pay gap between the two locations.

This matching of wages helps to equalize the prices of goods made in more than one place, and the profit rates of capital deployed there. And by confining labour movement within national boundaries, trade avoids social problems on both sides of a border that gets overstepped. Because migrants tend to be better-trained and brighter-minded than those they leave behind, their departure drains cohesion and efficiency from the community of origin. Because they stretch the cultural differences and squeeze the pay differentials of those they join, their arrival similarly disrupts the destination community. In contrast, say the free-traders, by transacting at a distance, through the goods that they produce, disparate people can sidestep the friction that comes of living side by side and dealing face to face.

## The Distributional Dilemma

If trade is so good for all who take part in it, why are the barriers so hard to knock down? Why are nations so quick to re-impose import quotas or 'anti-dumping' tariffs when the growing gets tough? And why is the World Trade Organization (WTO), notionally neutral referee of the global exchange game, such a perennial cause of anti-globalization crowd trouble?

The problem is that 'gains from trade' are strictly top-down. Economists can only show that countries' *average* income rises when they specialize and start trading rather than trying to make everything for themselves. There is no guarantee that everyone within the country will be better off. In fact, the theory predicts that gains for people whose skills are relatively scarce in the trade-exposed sectors will be accompanied by losses for those whose skills are relatively plentiful. For richer economies with large capital accumulations, this means reductions in income – or, if these are resisted, loss of jobs – for workers in labour-intensive industries with less than the average capital to work with.

This has been painfully true for richer economies in recent years, with employment in textile, basic industrial assembly and mining sectors devastated since these were stripped of the last of their protections and 'voluntary restraints'. Lower-income economies have seen the opposite reversal of fortunes. Workers in capital-intensive industries suffer job and pay cuts when their plants' 'infant industry' protections are dismantled, while labour-intensive commodity and agriculture labour is left better off.

Economists can continue to label free trade as a force for good, because total national income is increased. The winners from tariff removal could therefore compensate the losers for their pay cut, and still be better off themselves. But economists are equally clear that such redistribution reduces efficiency and so cuts total income – because taxes and benefits loosen the link between prices and costs, and weaken the incentive to make money by exploiting the gap between them, so actions that could improve resource allocation no longer take place.

Conventional trade theory promises that, in time, workers whose sectors prove not to be competitive will find better-paid opportunities in those that are. But such redeployment usually depends on moving to a new place and acquiring a new skill, the costs of which must once again be privately absorbed. So those who stand to lose from a move to free trade, with no compensation for their sacrifice, have every incentive to stop it happening. Trade barriers become a way of boosting demand for labour, so raising its wage, in less competitive activities. Since these are often once-great industries whose lobbying power outlasts their profitability, free trade is frequently frustrated by the combined efforts of 'old economy' unions and management to stop the imports coming in.

## Unequal Exchange

The problem of unbalanced distribution of the gains from trade goes much wider than unreconciled winners and losers within one country. Beneath the world-scale benefits, whole countries can lose out when protective barriers are lifted. Traditional trade theory says only that combined national income goes up when nations enter a free-trade arrangement. It does not guarantee an equal share-out of the gains. Protectionism can still boost one country's income, even if it robs all the rest. And as international income redistribution mechanisms are even weaker than those at national level, free-traders have few options to bribe the renegades not to put their barriers back.

Protectionism's attractions grow when a nation's current competitive strengths aren't those that it wishes to base its economic future on. Free trade forces specialization in areas of immediate advantage. But a one-off boost to national income from adopting free trade gives no assurance of continued gains through faster economic growth. Specialization in its currently most competitive sectors can prevent a country from adopting activities which would be competitive in future, or which enjoy the strongest rise in international demand and price.

Western Europe and North America did not get rich by specializing in the farm-based activities that their nations were built on. Nor did Japan, when it transplanted their first-mover advantage to Asia a century later. These nations' developmental master-stroke was to re-invest resources from agriculture into industrial production. The switch runs high risks if misdirected. Whereas unwisely purchased cattle or crops can quickly be slaughtered for a feast or ploughed back into the field, factories and machines have little value if their outputs don't find buyers. But if the risks pay off, the rewards keep flowing. Success breeds success, through a variety of self-reinforcements which cyberneticists term 'positive feedback' and economists prefer to call 'increasing returns to scale'.

Once industry gets going through the re-investment of agricultural surplus in manufacturing, it fuels its own advance. Expansion of output lowers costs, which raises industry's profits, which raises the finance for a further round of expansion. On-going investment also encourages innovation, which raises productivity and profit rates again. Growth of supply is sustainable because rising productivity allows higher wages, which – along with the demand for investment goods – ensures that all of the extra output can be sold. Clusters of industry can sweeten the pitch for further development by helping companies to cooperate on building communication and utility infrastructures, upgrading workforce education and training, exchanging information, and paying taxes for governments to provide more of the same.

Supply creates its own supply. One reason why industry goes where industry has already gone is because the skills, components and supporting facilities are already there. Evangelists may see merit to building cathedrals in the desert, but industrialists prefer to put their factories in the jungle of a ready-grown commercial district. To strengthen this centripetal tendency, supply (at a micro level) creates its own demand. Investors find their added output snapped up, and innovators their new ideas seized on, because there are businesses on hand eager to put new techniques and trimmings into their product, and showcase-hardened consumers waiting in the shops to buy whatever's new at the end of the line.

Increasing returns may also have been crucial in deciding why the first industrial regions (North America, Western Europe and Japan) remain the most advanced, despite the technological and institutional baggage that goes with growing up in the ages of iron, coal and button-down collars. And why regions that tried to industrialize later – despite all the hopes of leapfrogging old technologies and sidestepping outmoded institutions – have generally fought and failed to catch up.

## Cold Comfort Commodities

Industrialization's virtuous circle has a second lead-stretching mechanism that kicks the latecomers when they're already down. Rich countries have specialized in the range of goods and services which, as well as having increasing returns in production, tend to be in increasing demand. As buyers' incomes rise, their purchase of cars, electronic goods, leisure products and other 'sunrise' goods and services increases more than proportionally. These items' prices rise relative to other, less favoured products; or, if prices fall, the impact on sales is more than offset by a rise in demand. The rest of the world is left producing older items (especially farmed and mined raw materials, mechanical and basic electrical products) whose demand stagnates or falls as income rises, so their relative price and sales value steadily sinks. For countries trading internationally, this means a steady deterioration in the 'terms of trade'. Increasing volumes of their own output have to be exported to maintain the same volume of imports.

The products that trap a trading nation in this downward spiral used to be primary commodities. Raw food, raw minerals and other natural inputs tend to be in relatively fixed demand, so that any increase in supply forces a sharp fall in price if stockpiles are not to build up. Dips in demand, caused by recession in the consuming industry or resource-importing economies in general, can also push down prices and wipe out commodity exporters' profits, which in turn restrains them from investing in downstream activities with greater growth potential.

A favourite anti-globalization tale of woe concerns the farming family which used to grow all its own food, swapping the surplus for those goods and services that they chose not to self-provide. The risks of doing so were low, because all they needed was locally provided. The rewards were sufficiently high to keep investing in new machinery and land management, so they could raise their yields without damaging their soils. A rising population could sustain decent living standards, social integrity and natural harmony – until free-trade temptations intervened. Now, having scrapped their subsistence harvest in favour of a 'cash crop', they rely on an income which is permanently depressed by the dumping of subsidized industrial-country farm products on world markets, which swings erratically because of speculative global market movements, and which is in long-term decline against the price of other goods that they have to buy. As global market pressure takes an axe to their incomes, the impoverished peasants must take an axe to their trees – and dam their rivers and throw chemicals on their soil in a desperate effort to squeeze more out of it, setting up an ecological disaster to add to the economic trap which mistaken specialization has already sprung.

Commodity status is, however, no longer confined to farm,

mine, plantation and fishing ground. A growing number of manufactured goods, cast-off by their rich-world pioneers, has joined the commodity category. Basic computers and microchips, generic chemicals and drugs, clothing and shoes, data-processing and call centres are among former growth areas now dropping from premium to giveaway price – and relocating from geographical core to periphery as they do so.

To mainstream economists, this is a technology transfer which should dispel the idea that early trade opening locks nations into inferior specialization patterns. Industries go to where wages are low, but the effect of the move will be to close the international wage gap. If a developing nation acquires the same technology as richer rivals without wages rising to match, it is because their productivity remains lower. The inequality will be corrected once they learn to use the new machinery to greatest effect. But to free-trade sceptics, the refusal to pay people as much as their rich-country counterparts is another form of trade-based exploitation. Treaty-approved theft is transposed to the international level. The wealthy world can stop robbing its own labour – which might rebel – and instead short-change a more distant workforce, whose protests are easier to crush.

Commercial first-movers retain their technological lead through a concentration of industrial research and development which is even more extreme than their concentration of industrial activity. Whether measured by patent filings or Nobel Prize awards, serious innovation is confined to a handful of countries which can concentrate profitable technology-based companies, research-based universities, and big enough salary cheques to attract top scientists and engineers from all over the world. Where companies have set up R&D facilities abroad, their purpose has generally been to adapt existing products and processes to local conditions, not to devise or develop anything significantly new.

## Interruption of Service

Services were ignored in the first three rounds of post-Second World War trade liberalization talks because they were viewed as essentially untradeable. Some were marketable – including accounting, banking, insurance, retailing, telecommunications, transport, energy supply – but had to be provided at the point of use. Direct investment, not export, was therefore the route to customers abroad. Other services, especially education, healthcare, housing and public transport, could not even be sold *in situ*, because of practical or political obstacles to charging any price for them. Items that couldn't be stored and transported, for sale at another time and place, were simply left off the early trade agenda.

Three changes have ended this exclusion. First, technical change has made more services marketable at a distance, without the need for a physical presence on the buyer's home patch. Insurance can be sold on the Internet, inquiries from Europe handled by call centres in India, consultancy advice conveyed by teleconference. Second, many 'non-market' services have been pushed into the commercial arena by governments deciding that profit-making prices can be charged for them. Private school chains across America, dental showrooms on British high streets, franchised mail delivery in Sweden are among recent fallout from government decisions that previous price caps can be withdrawn, given technologically-reduced production cost or people's increased willingness to pay a premium over it.

Third, even with public services still kept out of the market, a growing role is being found for sub-contracting to private management. In a sometimes spectacular gamble with the long-term health of public infrastructure and finance, governments from Budapest to Manila are letting private companies build hospitals, roads and schools. The amount that the public will eventually pay, to cover private contractors' higher capital costs and give them the promised returns, could vastly exceed the initial reduction in public expenditure. Such public–private partnerships represent a massive leap of faith by governments in the ability of private-sector management to save money on completing and subsequently managing the facility. And an equal leap of faith by private managers in governments' ability to resist revising its terms or re-nationalizing the asset if its shareholders are found to have reached too good a deal.

International trade in services raises new concerns about economic efficiency and social fairness, on top of all those related to trade in goods. Whereas the success of opening goods markets to trade is usually measured by how much cheaper things get, prices often rise when service sectors are opened. Private companies seek profit on operations which, unless monopolized as an extra source of tax, were generally loss-making beforehand. Success is mainly gauged by immediate savings to the public purse. Opening to trade in traditional public services – education, healthcare, housing, social welfare, police and prisons, electricity and gas, water and sewerage, road and rail – raises the question of how far user charges can be raised, before the public ethos of allocation on merit gives way to the private allocation on ability to pay. Anti-globalizers cherish the example of Bolivians filling the streets until the government reversed its plans to privatize their water supply. But in many other instances, like Britain with its railways, higher prices are welcomed by those with money – as filtering out less appealing fellow travellers and giving an excuse to go by car. Those priced out might protest, but not often loudly enough to disrupt the contract negotiations.

The opening of economies to financial service trade raises their exposure to economic instability: of prices as money supply trends grow more erratic, of expenditure as asset prices find a new dynamic, and of the exchange rate as capital inflows and outflows step up. Their opening to trade in foreign communication media ends a long televisual free lunch. Commercial television which wrests audiences from 'serious' public-service channels, and those whose pictures were once channelled cheaply by a monopoly public buyer, means that viewers must now take out a subscription, or pay-per-view.

These risks of restricted private access and disrupted economic activity explain why the General Agreement on Trade in Services (GATS), which emerged from the Uruguay round of world trade talks in 1994, incites much more hostility than the General Agreement on Tariffs and Trade (GATT) on which it built. GATS is seen by its critics as a plot by wealthy nations to infest the rest with its privatized welfare states, swamp their wires and airwaves with imported programming, and surrender their next public projects to foreign engineering firms. Where over-borrowed governments have faltered on debt repayments in the past, they often cannot afford to fund any more investment themselves, and have to call in the private partners. Given the recent record of such deals to date, it is no surprise that there are rather more tear gas clouds to report and broken paving stones to renew, when public procurement officers return to work after their latest GATS negotiation. In 2001, according to a survey by the World Development Movement, more than half of the unrest in emerging economies was linked to protest against the impact of 'adjustment' policies to keep them open to trade.

## Regional Blocks

Coalitions of sovereign states have been steadily growing, in both the number and the strength of ties between their members. These regional unions and associations typically begin with economic and cultural exchanges, but move towards political and regulatory integration. Some of the strongest involve nations which, for most of history, have been perennial enemies rather than allies, with territorial rivalry eclipsing shared interest in making the most of what comes out of the ground. The European Union (EU), North American Free Trade Area (Nafta) and Asia-Pacific Economic Cooperation (Apec) represent expanding blocs built around the world's three major economies. Their initial mission was to remove trade barriers, stabilize exchange rates and let capital flow freely, but their remit has subsequently broadened to legal harmonization, migration, environmental regulation, competition policy, public purchasing, and other rule-setting areas which move beyond the economy into justice, social, cultural and constitutional policy. The EU has gone further, creating a framework for social policy, foreign policy, defence and institutional integration which many critics (and some champions) view as the vanguard of a single European state.

Regional cooperation is often viewed as a stepping stone towards global cooperation: a chance to try out cross-border integration on a small scale, with geographically and historically close political units. Supporters of the step-by-step approach hope that successful linkage at the regional level will make people want to see the process applied across a wider geo-political divide – and give them a short-cut to doing so, by building their one world out of intermediate blocks.

Grand production designs can sometimes be achieved by building separate mechanisms and then welding them together. Grand planetary designs aren't normally attained in this way. Cars, shoes and space stations tend to take shape more quickly and last longer when built by piecemeal methods. But worlds fashioned in this way tend quickly to fall apart. That's because mechanical systems never amount to more than the sum of their parts. How the finished assembly behaves is fully explicable from how individual components behave. Garbage in explains garbage out, and roulette-wheel computations can be traced to a crashed chip, however much the machine may seem to have developed a mind of its own. Social and political systems, in contrast, experience feedbacks and interdependencies which mean that the whole doesn't just exceed the sum of the parts, but can be fundamentally different in appearance and operation.

The last serious attempt to build a world system on regional sub-assemblies was the League of Nations, founded in the aftermath of the First World War and buried by the regional re-fragmentation that gave rise to the Second. The League needed to assemble only a small number of European and American nations, because they had the rest of the world fastened under their imperial belts. It took another world war to show up the fundamental flaw in the piecemeal system. The League's successor, the United Nations (UN), reverted to individual nations as building blocks. Many were first assisted to escape the imperial yoke before forging new cross-border links, so that top-down colonial unity could give way to a genuinely bottom-up international community.

'Global' governance institutions are relatively powerless today, because regionalism has taken root again. Being too far above sovereign nations' heads to have meaningful powers or funds entrusted to them, global agencies like the United Nations, World Trade Organization (WTO), World Bank (WB) and International Monetary Fund (IMF) are routinely wrong-footed when political conflicts or economic crises strike. The UN stopped only one war (the Korean) in its first 60 years; at most other times, its blue berets are forced to look on while national wars are fought under its name, or civil wars under its nose. The WTO's under-resourced secretariat takes months to investigate unfair-trade complaints, and has only limited powers to enforce its judgements. The WB extends its long-term development loans only after having checked the borrower's loan-worthiness with the IMF. The IMF is still scrabbling around for the 'adjustment' formula that will get creditor countries off its over-stretched books – and for the capital boost, repeatedly refused by its rich-country shareholders, that a genuinely effective set of turn-around policies is likely to demand.

The power that these agencies lack has risen to the supra-national level, pooled by like-minded nations into such exclusive alliances as the North Atlantic Treaty Organization (Nato) for Western-centred defence, the Group of Eight (G8) for big-power political coordination and the Organization for Economic

Cooperation and Development (OECD) for rich-nation commercial integration. These intermediate formations block the passage of power to the genuinely multilateral level.

The more successfully parts of the world link themselves into alliances, the less likely they are to knit one planet out of the patchwork. Economists affirm this by showing that there is an optimum span for single-currency areas, which only work between nations whose economic structures and behaviour patterns are comparatively close. This nearness is needed to ensure that members suffer the same economic 'shocks' at around the same time, and react to a common shock in similar ways. Shocks in this context are disturbances which, negatively or positively, affect the whole economy, or sectors sufficiently big and interlinked with it to have an impact on the whole. Internal shocks might arise from such domestic developments as an outbreak of strike activity and high wage demands (both common occurrences when unemployment gets low), the opening or closing of a major industry, a stock market crash, or a sharp change of political direction. External shocks arrive most often in the form of sudden changes in price of a key import or export (such as oil), war or natural disaster (with its disruption to markets and supply sources), or boom or bust in the economy of a major trading partner.

Policy changes are the alterations to interest rates, taxes and subsidies, government budget deficits, exchange rates, trade protections, industrial support, and other instruments that influence the level and pattern of national output or expenditure. Only when the policy adjustment needed to contain the shock is roughly the same for each constituent nation can they safely submit to a common set of policies. This is especially true of adopting a common currency, since this takes away the right to vary the national exchange rate, which adjusts all of a nation's wages and prices in relation to others', and so has long been the most powerful means for restoring the balance of trade and growth after an asymmetric shock.

Political economists add that there is also an optimum size for governmental structures. As governments grow, they amass the power to do more things to (and occasionally for) more people. But their politicians lose the power, or political will, to discern what people want done to them. And their public servants lose the power, or bureaucratic will, to do exactly what politicians tell them. Dis-economies of scale set limits both to governments' width (the number of citizens that they seek to represent) and to their height (the number of levels at which decisions are taken on behalf of those citizens). As a political unit's territorial span grows, it has increasing scope for the co-ordination of activity, and management and redistribution of buying power, that bind society together. But the people who feel short-changed by the system – paying in more than they take out, or doing more for others than others do for them – have equally increasing scope for short-circuiting the system. They can start enjoying the benefits of coordination and collective provision without paying their share of its costs, because those in power can't look everywhere, so don't detect and punish their 'free ride'.

A political unit's efficiency can also grow as its hierarchy grows higher. This gives more room for detailed division of decision-making labour, as more people share in the processing of information and implementation of actions that result. The more assistants that a minister has, and the more secretarial and number-crunching support that those assistants can draw on, the faster – in principle – can a well-reasoned choice be arrived at. The introduction of subordinates, and subordinates of subordinates, also helps to widen the span of government influence, and so spread its costs. So higher and wider government attains more decision-making efficiency (cost per person of making the decision that suits the most people), as well as more efficacy (the number who have an input to the decision and are affected by it). But the state's vertical growth, like its horizontal spread, is self-limiting, because extra height brings matching disadvantages. Politicians too far above the common ground aren't fast enough to see what's happening there, so can't move quickly enough when something needs to change. They're still steering the ship of state in one direction when its crew members decide to try another, still making triumphant speeches when the crowd gets bored and starts to boo and hiss. At a certain point, the growth of a nation state, onwards and upwards, stops making it stronger and starts making it shakier. Different forms of government have hit this limit at different widths and heights, but all hit it well before they can spread out or look down across the whole world. Empires couldn't spread across the whole world, and neither can the newer regional alliances. As forces of attraction gather strength within small clusters, so do the forces of repulsion between one cluster and the next.

## Getting Vocal for 'Glocal'

Far from being stepping stones to one world, regional groupings swiftly turn into stumbling blocks. Globalization and localization aren't opposites at all, but two parts of the same formula for constructive use of power. Regionalism, the force that drives them apart, becomes the enemy of both. This is because globalization aims to transcend the national state, dispersing its powers to different levels where they can be wielded more effectively and fairly. In contrast, regionalism aims to extend the national state, concentrating its powers at a middle level which is frequently malign. Stuck between the individual and the collective, regional authority tends to be too low to see and serve the general interest, but too high to give all personal interests the respect they deserve.

The European Union's eternal tensions reflect its troubled birth as a regional grouping of member states with globalizing tendencies. On economic issues, EU statespeople would like to be citizens of the world. They dream of linking markets, perhaps even monies, with equivalent customs unions centred around the US and Japan, and doing their bit for world development and defence efforts through the UN. But when attention shifts to politics, they revert to one of Europe's 15 voices, discovering that lands south of Spain, east of Romania or north of Poland are just too distinct to be part of the common home. As with most European projects, results to date have been the reverse of what was intended. Early economic integration efforts made the EU an exclusive regional bloc ('Fortress Europe') which stood in the way of wider trade and capital movement, and set off a still-simmering protectionist battle against the US. Later political and military integration efforts, aimed at strengthening regional identity, have been gallingly globalizing to convinced Europeanists – with the UK, France, Germany, Spain and other former imperial powers finding much more in common with ex-colonies outside Europe than with old enemies they were supposed to embrace within.

When the nation state is peacefully caught between localization and globalization, sovereignty is freed to find its natural level. With central governments' monopoly broken, power can flow upwards or downwards into the hands of those best placed to use it. Globalism is consistent with 'subsidiarity', the principle that power starts with the people. It should only rise above them if they benefit by having it exercised over their heads; and it should stop rising when it reaches the lowest administrative level at which the process of decision can be fully inclusive, and the spillovers from resulting action resolved by agreement among all those whom they affect. Globalizers disagree on precisely which powers, in an interconnected and interdependent world, need to rise from national to global level. Some believe that tasks like setting and collecting taxes, regulating enterprises' conduct and competition, redistributing income between people and places, and coordinating economies to avoid inflation and unemployment, can still be adequately done by national governments. Others argue that some or all of these must now be taken up to world level. The need to debate and decide these allocations democratically makes it a slow and often chaotic process, which is why world trade negotiations have been going on for more than half a century, and why world government has been on the drawing board for at least a hundred years.

But when the nation state is aggressively hunted by forces of regionalization, sovereignty gets trapped at artificial levels where self-interested rulers can abuse it. Members of a regional bloc try to build their togetherness on similarities among themselves, and getting together to attack a common problem. But the common home soon makes them behave like a tightly-knit family: the closer people get, the more often they tread on others' toes and lose their own tempers. So regional members inevitably try to reinforce their togetherness by dwelling on their differences from others, and getting together to attack a common enemy.

With deeper interconnections setting up wider interactions between people, companies and governments, rule-makers and referees have to climb higher to stay in command of the game. A question now facing the world is whether to let them climb half-way up and run world affairs through an assembly of regions, or to let them climb all the way and try to take decisions that affect everyone right from the top. The deepest dilemma is not between local and global, but between their world-embracing 'glocal' synthesis and a regional reaction which is antithesis to both.

The present approach to integration, through growing trade and capital market links, puts us on the fast track to regionalization. The Nato military alliance, EU economic alliance and US political alliance have shown themselves quick to bring together those partners who are geographically and strategically close, but incapable of reaching out beyond, to those perceived as more distant or different. Instead of creating more friends that they can draw in, they conjure up enemies that they can keep out. So Nato aligned itself firstly against the communist Warsaw Pact, now against a terrorist 'axis of evil'. Europe and its euro-zone sets itself up against America and its dollar-zone, with which it now fights running battles – ranging from protection of international trade which the other side wants freed, to the protection of international criminals which the other side wants jailed. The US once linked with the European Union for a common front against what was then the Soviet Union. Now its allegiance is shifting to the North American Free Trade Area and the Free Trade Area of the Americas, with the EU and its former Soviet-bloc allies on the other side. The continuous sparring and constant shifting of regional alliances was well anticipated by George Orwell, and despairingly dated *Nineteen Eighty-four*. It is no less sinister for arriving two decades late.

An alternative approach tries to build global unity not out of the knotted hearts of nation states, but over their heads. Genuinely international agencies are needed to take responsibility for such issues as financial regulation, pollution control, aggregate demand management and defence. These must include all nations and give each of them a vote, with the disproportionate political weight thereby entrusted to the smallest deliberately used as a counter to larger nations' disproportionate economic weight. The world has been able to form these agencies only after moments of global upheaval. So those that we have now stem almost exclusively from the Second World War. When the subsequent war stayed cold, regional groupings rose again, quickly wresting power from 1940s creations like the UN, IMF, WB and WTO.

The inclusive form of globalization is a natural and necessary response to the big wide world's graduation to global village. It cannot be built successfully on old multinational companies, which were still national companies with overseas extensions to lengthen leverage and strengthen privilege back home. Nor can it be built on old multilateral institutions if, as now, they stay under the sway of big corporations and regional governments. Far from making it easier to bring all parts together, the formation of regional groups bars the world stage from its final show of unity. Resurgent regionalisms may be the first fruit of an internationalized economy and network society. Only a counter-acting globalism can make this economy and society last, by putting the widest interests first. The world can get back on its unifying track, and stop its coalescing parts from returning to a warring whole, only by breathing new life into those multi-lateral remnants of the last world war – or waiting to build new ones from the embers of the next.

## The Unenviable Record

The 'gains from trade' were always based on a special set of assumptions about the economies that engage in it. These include perfect competition among domestic companies (so that prices stay closely tied to production costs); constant returns to scale (so that an industry can halve or double production with no change in unit cost); immediate and costless mobility of capital and labour between sectors; and the complete information about present and future production possibilities and consumption plans which was later recognized as a necessary counterpart of perfect competition. If any of these conditions is absent, even mainstream economists are forced to concede that protectionism can pay – for the country that imposes it, and sometimes for the whole world. Most industries are now dominated, at national and even world level, by a few large companies whose relations are very far from perfectly competitive, and most reap significant economies of scale. In consequence, the list of exceptions promising gains from 'strategic' protection is getting uncomfortably long to prove the free-trade rule.

As if to confirm the sceptics' point of view, the past 40 years' negotiated reduction in trade barriers and growth of international trade has brought widening income and wealth inequality, both within and between the nations most deeply engaged in it. Because most of the increased exchange has been of industrial products, trade growth has been concentrated among the world's richest nations, represented by the Organization for

Economic Cooperation and Development (OECD). Even within the OECD, trade is largely contained within three large regional blocs – centred on the US, the EU and Japan – with relatively small trade linkages between them. EU member states export goods worth close to 20% of their total output to themselves, but less than 5% to all other parts of the world. The North American Free Trade Area – the US, Canada and Mexico – internalizes trade on a similar scale, being almost as de-linked from Latin America as from Europe.

This concentration reflects the 'two-way' nature of most post-war trade growth. The big increase in flows has been between rich countries selling one type of car or computer while buying others, or exporting industrial components that are later re-imported as finished goods. Traditional 'one-way' trades – of one distinct good for another, or raw materials for finished products – have generally stagnated or declined, as industrial nations find at least one niche in each big industry, and industrializing nations learn to process raw commodities before shipping them out. But while this step has allowed a small number of later developers (mostly in East Asia) to trade their way up as the theory predicted, the record of 'outward-looking' development strategies is remarkably patchy. From 1960 to 1980, when global protection levels were still high, both industrialized and developing countries attained average annual growth rates above 3%. In the 20 years since, as trade barriers came down, the rich world's growth rate dropped to little over 2%. Developing countries' average halved to 1.5%. Without the take-off by the East Asian 'tigers', this deterioration would have been even deeper. Growth in Latin America, restrained by successive debt crises, slumped to less than 0.5%, and the average African economy actually shrank.

## Promising the Earth

Poor delivery on the free-trade promise is at the root of present anti-globalization fervour. Protesters condemn the WTO for ignoring developing countries' need to nurture 'infant industries' before exposing them to trade. They attack the International Monetary Fund (IMF) for imposing conditions on foreign loans which prevent governments from using them to build up internationally competitive non-commodity industries. They despair of a World Bank for making its loans conditional on IMF approval – and whose use of appraisal methods that prioritize financial over social returns gives it a predilection for dams and roads which modify the ecology more than they magnify the economy, while shying away from public health, education and transport investments, despite their rates of return routinely beating those of private investments even when social returns are ignored.

Free-trade champions confronted with this challenge find attack the best form of defence. Their explanation for poor recent experience is not that open-economy strategies failed, but that they haven't yet been tried. New models of economic geography, drawing on new views of growth, trace the problem to halting the liberalizing bandwagon at a halfway house. While first moves towards free trade reward the winners and further squeeze the losers, further barrier removal can turn the transactional tables. Stopping the process here, far from rescuing the lower-income economies, will lock in the privileges of the rich.

The first key change that economic geographers make to traditional models is to acknowledge 'economies of scale' in most types of industry, so that a rise in output cuts production cost and boosts competitiveness. Scale economies mean that any nation which starts with a small advantage in the industrialization process – because it has cheaper local material supplies, a larger population and domestic market, or a head-start in a certain technology – can quickly open up a big lead. Lower unit costs enter a virtuous circle with rising demand, because greater competitiveness drives up export sales, and higher productivity allows higher wages, which drive up domestic sales. A small number of companies can thereby rise to dominate the industry within each nation, and a small number of countries can rise to dominate the industry on a global scale.

In a world in which technologies are simple – converting natural raw materials directly into finished manufactured products – scale economies (or 'increasing returns') are a recipe for the early industrializers establishing a permanent lead over those who start out in their shadow. But a second key change to the traditional assumptions, acknowledging a lengthening of industrial supply chains, turns the picture of perpetual privilege on its head. When transport costs first fall far enough to enable once self-sufficient nations to start trading, the initial effect is for industries to concentrate in the nations which have the biggest markets or were first to exploit important technologies. Industrialization's virtuous circle is confined to this core. The periphery loses its industrial grip, being forced to specialize in pre-industrial mining and agriculture, and its average income falls (compared with its own past as well as with the manufacturing nations that have muscled their way through).

So in its initial stages, free trade seems fated to give to those who already have. But as transport costs fall further, companies which at first kept all operations close to their high-income home markets find it cost-effective to move some supply-chain stages abroad. Especially ripe for relocation are the intermediate stages – raw material processing, crafting of non-mass-producible components, assembly – which tend to be most labour-intensive, hence offer the biggest cost savings when moved from high-wage core to low-wage periphery. In making this move, companies keep their more capital-intensive final goods production in their high-income home country, and continue to concentrate sales there. But by investing in production abroad, they bring the seeds of industrialization to lower-wage areas. This belatedly fulfils the economic prediction that capital flows from richer to poorer regions in pursuit of higher returns. Formal models show that it also reverses previously regressive redistribution, reducing average wages in core economies and raising those in the periphery. As intermediate production is outsourced to cheaper areas, richer economies' loss of relatively unskilled jobs (or wage cuts needed to preserve them against cheap imports) cuts average income in the industrialized nations. Poorer economies' creation of relatively skilled jobs correspondingly raises average income in the newly industrializing nations.

These initial demonstrations of how trade's dynamics could switch from sucking-in to sharing-out relied on a continuous reduction in transport costs. But this is a shorthand for other impediments to free trade that raise its costs in the same way as transport. Local taxes, regulations, bureaucratic inefficiencies and corruption, incompatible business practices and insurmountable language differences are among the 'transaction costs' that can keep the benefits of trade flowing centripetally, to its richer originators, instead of centrifugally to its poorer recipients. Turning the chill of internationalization into the warmer glow of globalization requires these transaction costs to be cut down just as forcefully as transport costs.

When the World Trade Center was pulverized on 9-11, the first shudder that ran through the Western world was one of humanitarian concern. But it was quickly followed, especially in business circles, by the spectre of transaction cost. By necessitating time-consuming airport checks, forcing up travel insurance fees, and making managers and sales reps watch their step abroad, terrorists had thrown a large handful of sand into the wheels of global commerce. Globalization's wider agenda – of privatization, deregulation and general government minimization – is powerfully reinforced by the promise of breakthrough benefits when trade gets closer to being genuinely free. A fear that one day's skyscraper demolition had succeeded in damaging this ideal where three years' anti-capitalist activism had failed is among the more awkward motives in America's post-Ground Zero anti-terrorist campaign.

The production of intermediate products on the way to final

products, and final production's tendency to grow faster than its inputs, were previously a troublesome complication for free-trade theory. Now they have been harnessed to redeem its ancient promise of rising global output combined with fairer distribution of the income that results. Trade gains turn out to inhabit the economists' favourite 'U-shaped curve', giving pictorial respectability to the perennial hope that every dive precedes ascent to even greater heights. Instead of stopping at the base of the U, we need to build up momentum for the next ascent. Present pain can even serve as a predictor of future gain.

The dangers of premature protest have often been highlighted at times of rapid change. If only the Luddites had accepted that jobs lost to new machinery would be replaced by new jobs designing and tending that machinery, they would not have taken pitchforks to their spinning-frames. If only Leninists had realized that starvation wages and slum housing were temporary traumas at the start of the great industrialization, they would not have commandeered the factories and thrown a planner into capitalism's works. From the alienation that a person suffers when their workplace is reorganized to the cost changes that a nation suffers when its currency is devalued, initial shock can give way to later rejoicing if allowed to play out. Free-marketeers' message to anti-globalizers is to put away the banners, silence the chants, and just be patient until free trade comes back to finish what it started. Sending the World Trade Organization the way of the World Trade Center would freeze the process with all the gains on the already-rich side, the patient poor's rewards being still to come.

## The Capital Connection

But globalization 'hawks' do not stop at asking for the bonfire of controls to continue. There is another strand to traditional trade ideas that they are only too keen to modify. This is the economists' traditional assumption that capital, like labour, need not move between economies, because the free flow of goods will level up its rewards without its leaving home. While free trade may help economies to speed the expansion of the industries they have, free capital movement can spark the arrival of whole new industries. Even as trade growth was outpacing national income growth in the post-war period, cross-border capital flows were outpacing them both. Far from being an alternative to trade, as earlier assessments often suggested, capital flow is the missing ingredient that might finally make it work.

The anti-globalization movement aims its harshest words and gestures at the newest targets for trade liberalization – financial services, public infrastructure, and intellectual property rights. It is here that movement of commodities shades into the movement of capital. So it is here that the clash of views is strongest, between pro-globalists who believe that the new phase of liberalization will heal the wounds of previous rounds, and protesters who see the knives receiving a cruel new twist. While the debate has reached a new intensity, it has also returned to older territory. For it is with the rights and wrongs of capital movement that the fight for world development began, more than a century ago. And with this return comes an even crueller twist. The battle lines are neatly reversed, with today's protesters taking arms against a process which on their own fair-trade logic should be viewed as their biggest hope.

## CHAPTER THREE

# Capital Punishment

*Global growth prospects are transformed when capital, and the technology and skills that it conveys, flow from where it's plentiful to where it's scarce. Nations can quicken their growth process and even leapfrog the trade partners that started out ahead. The trouble is, capital rarely flows to where it's most needed. When it does, the policies needed to attract and retain it often destroy more capacity than it's capable of building. But as with trade, better managing of the flow is a more helpful response than trying to choke it off.*

The world becomes a different place when capital starts to move. While free trade is praised on the political right and questioned on the left, free capital movement splits opinion on both sides. Karl Marx was a fan, and praised the modernizing exploits of expatriate capital far more eloquently than most of today's commercial cheerleaders. Many political conservatives are foes, decrying roving capital's commodification of culture as much as its destruction of economies at home and abroad. Many more regard the global flow as positive in principle, but are constantly caught out by the results.

When handled well, this magic import causes growth rates to take off, deserts to sprout high-technology factories, farmers' sons to bash metal and their granddaughters to manage boutiques. But badly managed capital inflow can tip nations into debt traps, make their currencies collapse, turn their old industrial heartlands back into wilderness, and help baroque technologies stamp on classical tradition. Whether change is for better or worse depends on whose capital flows, where it goes, and what it gets up to on arrival. The progress of a corporate-driven globe rests less on whether trade flows in the right direction than on whether the capital is going where its presence does most good.

## Calling on Capital

Anything counts as capital which helps to make a product without being used up or disappearing into that product. Factories and machines, 'physical capital', were the first and are still the most visible examples of goods designed for making other goods. The development of banking and shareholding led to the appearance of 'financial capital', sums of money deployed to make further sums of money. This established the general idea of capital being a stock, from whose operation consumable items flowed. People could start to accumulate capital once they ceased to consume all that they currently produced, and instead put some aside to help produce more in the next period.

On the day that they weave a net, coastal villagers don't catch any fish. As soon as they cast the net, the catch rises much higher than if they'd carried on with rods or spears. Factory hands produce no cloth while they're putting a spinning machine together, but much more than previously as soon as they switch it on. In general, capital investment is an up-front sacrifice that subsequently amplifies production and consumption. The greater the capital stock an economy has, the more production it can generate. Hence the more of a surplus over subsistence that it can raise, to invest in even more capital stock. Capital, and hence output, will keep on growing through re-investment until there are no more areas in which it can profitably raise production. This moment when prospective returns can no longer justify the associated costs and risks has long been predicted, but so far staved off. Periods when all requirements seem satisfied have repeatedly been followed by discovery of new wants and new ways to make people pay for having them fulfilled. New technology and new territory have been the perennial sources of a revitalized rate of profit.

Accumulation becomes self-sustaining, once an economy gets rich enough to set aside resources from being immediately consumed. But an economy too poor to start accumulating capital by itself needn't get stuck in the subsistence-level trap. Capital is itself a commodity, which can be bought and sold, or lent and borrowed, between people who want more and people who already have too much. Once capital markets develop, an economy can go beyond the painstaking build-up of capacity through incremental saving and re-investment. It can speed the growth by drawing on extra capital from spare pools of it, for which richer economies are struggling to find the most profitable use. What one group may have taken years or even centuries to accumulate can, if properly transplanted, be made instantly available to groups with shorter economic histories. Latecomers to the growth process can have the last laugh, by leapfrogging first-movers' preparations using the very resources that the profit motive drives them to supply.

Capital import can become a fast track to expansion. Once capital flows in from abroad, a nation's investment ceases to be held back by its domestic rate of saving. No longer constrained to save before its spends, a nation can launch whatever projects the international markets offer loans to, or buy shares in. The limit is now set by how much expansion will yield a profit, not how much can be financed from the current pool of saving. On traditional economic logic, capital should earn higher returns the scarcer it is relative to labour and other productive factors. So richer nations have, in principle, a good self-interested reason for investing in nations endowed with smaller capital stocks.

## Growing With the Flow

Capital import can rescue nations from a circular trap that might otherwise frustrate their solo efforts to go for growth. A closed economy can only invest more by consuming less. But people can't tighten their belts if they're already on the bread-line. To build up productive capacity for the future, people must avoid spending all that they've got right now. Yet such self-discipline can undermine their efforts to expand. Frugality leaves some of the current production unsold, causing the capital that made it to miss its payback target, so destroying the incentive to invest in any more.

Saving that creates the means for expansion in the long term destroys the incentive for it in the short term, because of the demand deficiency that results if people tighten their belts. In this 'paradox of thrift', virtue forestalls a virtuous circle, as a cash-strapped present blocks the route to a richer future. Once capital flows in, however, a nation can consume more and invest more at the same time. It can actually step up consumption, even importing more goods than it exports, because the investment funds arriving are a substitute for the spending power that's leaking into foreign goods.

Capital import can also be a short-cut to the technological cutting edge. When investment arrives as ready-made plant and machinery, or as equity and loans with advice on what to do with it, the capability to produce can be built up in tandem with the capacity to produce. By acquiring a technology later, new industrializers can go straight to its most recent version. With state-of-the-art facilities can come the latest product designs and process methods, overseen by the latest management techniques – any transmission delay often usefully serving to separate fact from fad. This gives later developers a chance to jump ahead of the pioneers who may be stuck with an earlier, less efficient stock of capital, and reluctant to splash out on the latest because it will hasten the redundancy of what's already there.

At first, foreign investors are likely to invest in export-oriented activities, using the 'host' country's lower wages to build cheaper products for higher-margin sale in the 'home' country's markets. But the new investment also enables the host country to start making for itself the goods it once imported; and to raise its labour productivity so that low-cost production doesn't force it to keep wages down. Rising domestic production is matched by rising income, so foreign capital also becomes interested in the host country as a market. This promises a further return for external investors, and strengthens their incentive to move in. If another country's producers want to sell services, as well as goods, to the new market, they have little choice but to send capital in its direction. Whereas manufactured items can be sent in from production sites abroad, services can't be stored and shipped this way. They need to be produced at the time, and in the place, that the buyer wants to consume them. So while it might be able to sell stereos, bulldozers or canned food from a distance, a multinational must set up on the client's home soil before marketing its financial, retail, restaurant or real estate experience.

Host countries, once persuaded that capital import can speed their development, can use various techniques to encourage its arrival. The 'carrot' consists of subsidies, tax concessions, investment guarantees, publicly-provided transport and communication infrastructure, fast planning approval, and other measures to reduce the risk or raise the return for capital that takes the cross-border plunge. The 'stick' is wielded by raising the tax on imports, until foreign companies can profitably sell to the local market if they actively produce there. These investment-promoting trade tariffs are generally accompanied by requirements to make the product with local inputs, to avoid an illusory import substitution in which final assembly takes place on the home patch but all of the components and skilled labour inputs still come from abroad.

The governments of capital-rich countries have also worked to promote capital flow to poorer neighbours: through direct (bilateral) aid and credit flows, and indirect (multilateral) flows via government-backed agencies like the IMF, the World Bank, the European Bank for Reconstruction and Development, and older Development Banks for Asia, Africa and Latin America. Aid is sometimes humanitarian, but usually recognizes a mutual interest in capital flowing from wealthy North to catching-up South. If it makes poorer countries a better customer or stronger production base for the donor country's multinationals, aid can generate a payback even though it's formally a one-way flow. This is why, despite protesting that it breaks banking rules and encourages a repeat of the reckless spending, governments eventually write off the debts that natural or man-made disasters have left poorer countries unable to repay.

## Wrong Direction

Financial indicators have a habit of going in the opposite direction to that predicted by economists. It is a habit that global capital flows were quick to pick up. Despite exhortations to go where current presence is smallest and prospective returns highest, most of the mobilized millions are flowing the other way. A flow of *capital income* from poor nations to rich should not be surprising. If they are to continue investing in 'emerging' markets for hard-headed commercial reasons, physical and financial capital investors must receive a worthwhile return. This is why host governments have been so keen to remove curbs on interest and dividend repatriation, and to stabilize their exchange rates so that local profits maintain their weight in dollars, euro and yen. Much more disturbing for economists is the tendency for *capital itself* – the 'principal' sum, as well as the income it generates – to flow from poor nations to rich. This happened in the past to inward-looking, import-substituting industrializers because they ran up such heavy debts that the amounts needed to service them eventually outweighed any investment flows still coming in. It has happened more recently to outward-looking, export-driven industrializers because local investors, once free to move their money, expected it to earn more in the pampered surroundings of a higher-income economy.

Economists have traditionally argued that this is the result of a distortion, wrought mainly by developing-country governments, whose over-regulation of private enterprise depresses the high returns that local capitalists should be able to enjoy. Their case seemed vindicated in the early 1990s, when a widespread reduction of tax and red tape on the emerging world's investors led to a sharp rise in inflows to it, both of foreign capital deciding that the rewards now outweighed the risks, and of domestic 'flight' capital sensing that it was safe to return. According to UN data, foreign direct investment (FDI) and long-term lending to developing countries rose to $655 billion in 1994–7. FDI inflows grew at twice the rate of trade flows in 1985–90, and three times that rate in 1990–4. Investment banks assembled teams of emerging-market analysts to pick out tongue-twisting shares in far-off countries, and credit rating agencies had to dust down their atlases to ascertain the worthiness – and whereabouts – of such new credits as Estonia, Peru and Vietnam.

But this turned out to be the high point of rich-world interest in the profit potential of the poor. By 1997, emerging markets were coming unstuck over artificially strong exchange rates, stagnating economies and unsustainable debts. Foreign bond and equity buyers retreated, and developing countries' long-term capital inflows dwindled to just $19 billion in 1998–2001.

The surge of funds across the wild frontier turned out to have been the result of two short-lived, and swiftly reversible, rich-world developments. One was an outbreak of technological innovation around cheap computing and the Internet, which whetted investors' appetite for exotic 'technology' stocks. These are viewed as closely related to emerging-market stocks, in terms of offering high potential reward in exchange for high risk, involving investment in a relatively unknown area, and being easily prone to herd behaviour by financial investors, whose mass movement into and out of emerging-world equities and credits has been known to treble or halve their value within a day. Instead of finding money for ambitious new plans to etch microchips onto headlice or sell binliners online, some investors were persuaded to sink similar sums into ambitious new locations. The advent of a professedly pro-business administration (appointed or even elected), an easing of exchange controls and announcement of a currency target against the dollar were often all it took to prepare the ground for a feeding frenzy over the local stock market, or rapturous reception for the billion-dollar bond. A second development that helped to push capital towards the territorial as well as the technological frontier was the unprecedented availability of funds, due to a worldwide monetary relaxation. Central banks surprised by the severity of the early 1990s downturn in North America and Western Europe expanded the money supply and pushed down interest rates. Previously dodgy projects, and previously off-limits places, now promised returns sufficiently far above the cost of funds to be worth the extra risk.

The sudden drying of capital inflow to developing countries coincided with a tightening of liquidity in the world's main currency zones, as faster demand growth put inflation targets in danger, and share prices rose so high that any further puff could make the bubble burst. Repeating a pattern which historians can trace back through the booming 1960s and 'roaring' 1920s to the first peak of modern globalization in the 1860s, the renewed rise in world interest rates was followed by retrenchment of exotic stocks. Thailand sparked the general East Asian crisis in 1997, Russia and South Korea joined the debt and currency 'correction' wave in 1998, Brazil narrowly escaped the same fate by devaluing its 'fixed' exchange rate in 2000, and Argentina hit the rocks by refusing such adjustment in 2001. In between these emerging-country asset falls, emerging technologies were also coming down to earth, with dot-com shares turning downwards from early 2000, and the telecom companies that channelled them finding their stock market calls cut off throughout 2001.

Because the peak flow of funds into emerging markets was so brief and so decisively reversed, its effects on the global stock of inward investment have been unspectacular so far. Foreign-held assets, a measure of how much capital has crossed between countries, surpassed their 1914 level (around a fifth of world output) in the early 1980s. But this partly reflects there being more borders to cross, with colonial assets being reclassified as foreign if their newly-freed hosts refrain from nationalization. The figure is also inflated by foreign financial holdings, whose permanence and productivity stop well short of those associated with the 'fixed' industrial assets which predominated before 1914.

Although successful FDI generates follow-up investment, most of this is financed by profit made and retained in the local economy, or shares and bonds sold to investors there. Once an operation is up and running, it can thus often fund its own expansion with no need for any more capital inflow. Follow-up commitment also averages only one-third of the original sum for developing countries, compared with more than two-thirds on FDI from one rich country to another. Whereas every $1 of profit flowing back to the rich-country source was offset by $3 of new FDI in 1991–7, the next four years saw substantially less passed over and more withdrawn. The balance of new long-term lending against interest and principal repayment shifted against lower-income economies in a parallel way.

Despite their efforts to attract it, few low-income economies have been big recipients of foreign capital. Most FDI still flows between the wealthiest economies. What moves outside them has been concentrated on a handful of new industrializers, mostly in East Asia and Latin America. Worldwide, only 20 countries could look to FDI for more than 20% of their fixed investment needs during the 1990s, and only a few of these found their export capacity significantly strengthened by it. Even these have, through repatriation of interest and dividends, ended up seeing much if not all of the original FDI inflow trickle out again.

Multinational companies' interest in the reverse flow of

capital was reviewed in the opening chapter. When enterprise steps abroad, it is usually to get a better grip on business at home. If MNC investment in other countries – whether as market or supply base – doesn't pay itself back within a realistic timescale, the cause is more often miscalculation than magnanimity. Fear that foreign investment would drain their stocks of natural capital and constrain their accumulation of complementary created capitals was precisely what drove later industrializers, from Germany to Japan, to keep exchange controls in place until they were rich enough to enter world capital markets as net suppliers. The requirement to open up stock and bond markets and banking sectors, as well as goods and service trade, at a much earlier phase of development, means that today's new industrializers must play by substantially different rules. Initial inflows of capital that later reversed themselves, as merchandise trade moved into surplus, would signal success for this early exposure. The more usual pattern, of initial outflow followed by dependence on inflows as trade moves into deficit, suggests that yesterday's rich have played a trick in not letting their history be repeated. Nations that start their modernization drive with borrowed or donated capital are left forever struggling to switch into growing their own.

The contraflow of capital from 'South' to 'North' doesn't only exacerbate global income and wealth inequalities. It may have been what started them off. There is a strong argument that seizure of capital by European nations from their conquests elsewhere in the world played a key role in giving them a head start in the growth race. Britain, the first industrial revolutionary, has long been accused of building its prosperity on poorer people's savings. From plundering South American gold to stripping India of its upstart textile industry, it was a motherland which placed firm limits on how much capital its colonies should keep.

On one charge – that it used foreign markets to offload unsold products and sidestep the 'paradox of thrift' – history probably acquits the British Empire. The rest of the world had little income to spare at this time, and the products that Britain was making the most of – metals, machinery, textiles, fuels – weren't the things that the world most obviously needed to spend it on. Captive colonial spenders couldn't fill the gap left by domestic savers. But the empire-builders found an equally good way around the demand deficiency trap by continuing to consume almost all that they produced, and grabbing the capital they needed from abroad. Some economic historians use this idea to explain how such strong 'take-off' could occur with a savings rate that stayed so low, compared to those run up by nations industrializing later. It is also what some historians in Britain's ex-colonies conclude, when they support demands for compensation for the capital that their unwanted visitors are said to have taken home.

## Divide and Rule

The rarity of useful ideas in economics means it would be wasteful to leave them trapped there. Once the capital concept was out of the bag, many others found it lurking on their territory. Sociologists discovered 'social capital', the stock of personal acquaintances and authority with which people can amplify their personal productivity. Educators reclassified their life's work as 'human capital', the skills and experience that let someone do a better job. Organization scholars drew attention to 'intellectual capital', the unique mix of social and human capital that helps companies and other agencies to go about their business. Sensing that the profit-making ground was being wrested from them by theorists who didn't know their differential calculus, economists struck back with 'knowledge capital'. This is the stock of scientific ideas and technological wisdom which is often bundled into human and knowledge capital, but can exist separately, doing in the computer age what turbines and traction engines had done in the age of steam.

Radical socialists, seeking to explain how inequality endured in societies in which stark class distinctions had been eroded, alighted on the idea of 'cultural capital', the access to education, training and privileged social contact through which they could quickly re-acquire control over physical and financial capital, even if communist tanks or high inheritance tax appeared to have stripped it all away and spread it around. Democratic socialists reinterpreted their campaign as one in defence of 'public capital', the physical infrastructures, social services, national insurances and business regulations which the government had to provide in order that private production could function, because they could not themselves be sustained through private production.

Not to be outdone, political conservatives pointed to the need for 'institutional capital', the social arrangements for upholding law and order, resolving disputes, enforcing moral rectitude in private affairs and preventing corruption in public affairs, which good business needed and which too much state intervention could quickly destroy. Outside the debating chambers, environmentalists found that the forests, wetlands and coral reefs they'd been defending were all forms of 'natural capital', whose flows were the clean air, water, food and ultra-violet protection without which other capitals would soon be robbed of any place to live.

## National Capital: Getting Personal, Going Portable

When political economy was still a sport played at national level, critics saw a class-based division within these various capitals. Some forms had to be physically embodied by the person who owned or controlled them. Others could be separated from the person without their being deprived of control. Physical, financial, public, institutional and natural capital line up on the 'disembodied' side. Human, cultural, intellectual and social capital are all partly or wholly 'embodied'.

The difference matters because acquiring capital always involves a risk. People can run short of things to live on right now, if they put too much aside for making more things to live on in future. And they can find themselves no better off in the future, if the capacity in which they invest turns out not to have any use. A machine has obvious construction costs, and a training course has inescapable fees, in addition to the immediate pleasures that must be set aside while these projects are undertaken. There may be no payback on these costs if the machine makes a product that no one wants to buy, or the course leads to a skill that no one wants to employ.

The risk of misplaced investment gets worse as skills and equipment get more specialized. This reduces the chance of finding another use for them – or re-selling them to someone else who can do so – if they prove to have no value in their original purpose. Risk also rises with the necessary minimum size of the investment, since more of the total capital then has to be tied up in one hit-or-miss application. These problems of asset 'specificity' and 'indivisibility' were especially severe for the industrial pioneers, many of whom went to the wall when their watermills were sidelined by the steam engine, or their timber forests made unsaleable because the shipwrights had switched to steel.

All this began to change when capitalists learnt to make their holdings liquid. Instead of owning physical stocks of plant and equipment, they hit on the idea of owning legal entitlements to the profit that flowed from those stocks. Because these 'shares' could be issued in their thousands, and held in small amounts, big holdings in one specific enterprise could be broken up into a portfolio of holdings in many, thus spreading their risk across a variety of technological and geographical areas. If one place or product bit the dust, there would be plenty of others still riding high. Shares were also much more easily tradeable than the factories and facilities that they ultimately represented. With the arrival of large 'stock markets', fortunes that once rode on a single enterprise could be spread across the whole economy, investors selling off the companies whose future didn't look good, and buying into those that had a bright one.

This paper escape, from ownership of fixed capital to ownership of legal titles to the income from fixed capital, didn't just let early capitalists reduce their risk. It also meant that they could cease to dirty their hands on machines, and focus instead on the figures that flowed from what they did. A sufficiently diversified share portfolio comprised stakes in more enterprises than an investor had time to visit. Their performance had to be monitored, to check that the shareholding was worth keeping. But with the development of accounting systems, a flow of numbers told the owners what they needed to know. How high a return their investment was making, and how much was retained for re-investment, could be gleaned without even going near it. In a further triumph for financial over mechanical engineering, lumpy loans were split up into convenient corporate bonds. Like shares, these could be held in a portfolio across many places and sectors, and readily re-traded, extending the escape of 'finance capitalism' from risky dependence on product and place.

This escape is, however, only possible for holders whose capital can be 'disembodied'. Those whose investment is in themselves – as human skill, knowledge, team spirit or social contacts – cannot just sit back and watch its returns flow in. To generate that return, they must still be physically present in the workplace. And because they cannot divide entitlement to their earnings into small units, selling them off to other shareholders, holders of human capital continue to have all their eggs in one precarious basket. A few gifted individuals may be able to build up a portfolio of skills, switching from coppersmith to nuclear technician to call centre supervisor as economic conditions change. Most have only a limited range of marketable skills, failure to sell which can lead to difficult phases of redundancy or re-training. Those who must carry their capital with them are at a basic disadvantage to those who can fold it up and lock it away in a document case.

This distinction points to a basic deception in recent arguments that the class divide is dead, on the grounds that everyone draws income from capital now. It is true that many employees now have access to significant human, cultural and intellectual capital, financed by state education systems and company training schemes as well as their own pockets. But getting the return on this capital means working as long and intensively as ever, at tasks far more demanding than financial investors' scan of closing price lists. Employees increasingly have a small stock of their own financial capital, as insurance and pension fund investments. But this is usually only a small fraction of their total income – too small to live on when they stop working, according to assessments in a growing number of countries. In Britain, where tax-financed state and salary-linked occupational pensions have been run down in favour of those linked to stock markets, poor performance at the turn of the century means that the average 26-year-old must now work into her seventies before retiring comfortably, according to 2002 calculations. But if too much basic pay rides on the stock market, even staying in work needn't solve the problem, as Enron workers with their pay linked to the company's share price found out only too well at the end of 2001.

Knowledge capital uneasily straddles the class divide. Some ideas are 'codifiable', so that anyone who listens or reads can glean the secret. Others are 'tacit', held by people who can't explain exactly what they know. Some social contacts and commands are personal to the enigmatic character who wields them; others attach to a title or office that can be passed from boss to boss. But for knowledge workers, the distinction is no help. The part that can't be separated from them forces them to stay full-time on the job. The part that can be separated generally is, and gets snapped up by others, in a re-run of the way in which early capitalists snatched the physical capital from workers who once owned their own materials and machines. Codification of knowledge is a recipe for de-skilling. Whereas the owner of a valuable machine can distance himself from it and still draw its income, the owner of a valuable mind can get away from its contents only by giving them away for someone else to profit from. This may explain why knowledge workers – especially in teaching, engineering, medicine and administration – were among the most militant groups as the era of 'within-one-country' capitalism drew to a close.

## International Capital: You Can't Take It With You

But just as this old class division was threatening to erupt, capitalism achieved one of the periodic shifts which wrong-foots its detractors. Globalization brings to the surface a new distinction between forms of capital. At issue now is its separation not from a person, but from a nation. Financial, human and knowledge capital can be easily transferred across borders. But the physical, social, intellectual, cultural, public, institutional and natural varieties all resist export to a considerable degree.

Financial capital can move abroad almost instantly, as electronic transfers among financial institutions. Human capital can move abroad through migration or expatriate working by its holder. Knowledge capital can cross borders by a combination of these methods. The codifiable component moves through electronic or printed transfers of information, and the non-codifiable component by people going abroad to tell others how to do the job, or to do it themselves.

In contrast, physical capital can go abroad only if initially sold for export, unless people are willing to disentangle it from an old workplace, move it, and re-install it somewhere else. This is rarely worth doing, because equipment installed some time ago will have deteriorated physically and been overtaken technically. A country's best hope of importing physical capital is to attract an inflow of financial capital, and hope that its investors will plough it into solid buildings and machinery, instead of just swapping it for share and debt certificates that they can easily re-sell.

Social and intellectual capital, which depend on network

links between members of a society or organization, cannot move around successfully unless the whole group is willing to uproot itself; even then, much is lost as the links are rearranged, and new ones added which lack the trust and talkativeness of the old. Cultural capital is often lost when borders are crossed, because being steeped in one country's culture makes it harder to understand or appreciate others'. Those who hoped that entertainment's pooling of national performance capabilities would bring the same hybrid vigour as economies' pooling of national production capabilities have long had to suffer the indignity of the Eurovision Song Contest, or the airport bookstall confined to those styles which improve in translation.

Because 'intellectual capital' is generally contained in the communication lines and knowledge stores of a company organization, new industrializers had hoped that hosting foreign companies would be a quick way to acquire it. But multinationals rarely move significant teams into newly-entered territory, even on temporary assignments or flying visits. And most are wary of trusting their more valuable intellectual capital to peripheral locations, fearing the high-fliers who 'go native' and won't come home, or the local pirate who ensures that vital information stays behind when they do. Inward investors, under pressure from host government as well as head office, tend to assemble new teams to handle the foreign project and employ local people to run it when the groundwork is laid. Unfamiliar with themselves and their surroundings, the new arrivals have to build up their own intellectual capital from the base.

Public and institutional capital have proved almost impossible to export, because they are bound up with particular forms of running and regulating a nation state. Formal statutes and structures of government can be easily replicated elsewhere, but not the many principles, procedures and informal understandings that go with them. Africa's ex-colonies found this in the 1960s, when their European-style parliaments and judiciaries failed to deliver European-style rule, generally sinking into corruption before slipping into dictatorship. Europe's ex-communist countries rediscovered it in the 1990s, when re-creation of the political and economic structures of their Western neighbours did nothing to kindle comparable economic dynamism. The letter of the law may be easily carried over, but its spirit is far harder to capture. 'Institutional quality' will remain a problem for newly-independent states long after dependence on a dictator or a colonist has ostensibly been shaken off.

Understandably, later-developing economies have tried to import the three types of capital that move: financial, human, and knowledge-based. They did not waste time on the other types, which resist transplantation and might take many years for a convincing imitation. Unfortunately, the three that travel can function only in conjunction with the seven that stay away. Worse, financial capital enters developing countries only if assured that it can exit quickly, an option that it exercises frequently, leaving serious economic vacuums in its wake. Human and knowledge-based capital enter only if skilled workers and researchers can be persuaded to move into the lower-income region. As a scan of the waiting lists for Northern-hemisphere 'green cards' and the enrolment lists at its universities makes clear, the predominant flow continues to go the other way.

Any capital that reaches low-income economies generally does so on an easy-come, easy-go basis. Foreign banks, portfolio investors and companies will commit those resources that they can be sure of quickly retrieving if the going gets rough. This lack of commitment shouldn't matter in principle, if capital doesn't do much when it arrives. But like a bull in a China shop, or Union Carbide in its Bhopal factory, a short visit can do a large amount of damage.

The fact that only three types of capital, out of a possible ten, can readily move across borders, deals a double blow to the strategy of externally-financed development. It means that the capital it imports will often do it no good, because it is only a subset of what is needed. And any capital it does attract will be a fair-weather friend, turning on its heels at the first sign of trouble. The power that once lay in separating capital's income flow from its physical stock now arises from the ability to move that stock across borders. Rich countries turn the investment taps, and make sure that they don't allow enough flow to let later industrializers completely turn the production tables.

## Natural Break

Natural capital is the exception to this maldistribution rule, being already widespread in the lower-income world. Nature renounced commercial prejudice in distributing natural resources. Since early industrializers tended to tap their own seams and chop down their own trees first, they have not yet had time or inclination to exhaust those that lie further south. Many of the biggest remaining natural resource deposits are thus held on or in the world's lowest-income lands. From Europe's extraction of South American gold to Japan's exploitation of Philippines forestry, richer countries' enterprises have responded by trying to move them to where they would be more commercially useful. Those that won't move, they have tried to pluck and process *in situ*, on the pretext of converting natural capital into other, more useful forms.

Undoubtedly, some of the felled forests now live on as the interiors of great buildings or good books, and some burnt oil is immortalized in roadways or industrial estates. The amount that was gratuitously burnt, with only the gas-cloud left for local consumption, is a moot point for Northern environmentalists and Southern economists. Without importing other forms of capital to build up processing industries, lower-income hosts of natural resources must generally export them for refinement elsewhere. This exposes them to the twin dangers of specializing in commodity exports of volatile and dwindling world market value, and of giving other countries control of their main capital stock in the process of converting it to other capital types. If a commodity holds its value and continues to be bought in large volume, the strong demand can inflate the exporting country's exchange rate, making other industries uncompetitive and undermining attempts to trade up from the primary base.

With all these risks arising from the exploitation of natural capital, its hosts could be forgiven for trying to leave it in the ground. Ways have recently been devised of getting financial rewards for doing so – notably the debt-for-nature swap, under which production-poor but vegetation-rich nations like Costa Rica get partial forgiveness from their creditors in return for a promise not to chop down too many trees. But such deals stray too close to green-tinged blackmail to win widespread lender support; and even where concluded, they cannot completely arrest the running-down of natural capital to repair holes in the balance sheet left by overdrawing on the artificial kind.

Recognizing that they cannot realistically transplant other countries' forests or pipe in their cleaner air, even when accepting these in place of debt repayment, the mass polluters have hit back at the other end of the natural supply chain. The emerging world may host most of the material still left in its raw state, but industrial nations are the major producers of what comes out of the chimney's other end. So they have made a pre-emptive grab for the world's landfill sites and 'carbon sinks' – those stretches of soil, sea and sky still clean enough to absorb more waste products. As with grazing land and other more traditional 'commons', it is first-come, first-served. Where deforestation has not yet left low-income nations with nothing to burn, this rich-world capture of the planet's pollutant-carrying capacity may well be leaving them with nothing to burn it in.

## Borrowed Time

Foreign direct investment (FDI) projects are planned in the head offices of the companies that bankroll them, sometimes with a walk-on part for the foreign aid agencies that top up their contribution. Governments on the receiving end can try to steer the FDI towards regions, sectors and local partners in line with their economic and social policy objectives. But there is a limit to the control they can exert, under the investor-protection laws required to make the funds flow; and to the control they want to exert, given that overseas firms feeling over-regulated in the new land can always go elsewhere.

Countries, like companies, that want to quicken their growth with outside capital tend therefore to borrow abroad, to supplement FDI if not to substitute it. Foreign debt, raised through bank loans or bond issues, can often be a cheaper source of capital than shares, especially when world interest rates and inflation are low, and when a country has natural resource stocks to secure its loans against. Unlike equity, debt can be issued by governments (backed by their ultimate ability to tax), and the proceeds can be used for any purpose, not the narrow range of tasks specified in a flotation prospectus or sanctioned by the paymasters who bought the shares.

The dangers of amassing foreign debt in pursuit of domestic development are regularly paraded on the international scene. Because the borrower remains in charge of where the money goes, debt finance is always especially popular with dictators, supreme leaders, juntas, national salvation committees, and other groups which seize control of government without having to capture people's electoral sympathies first. Autocrats' heightened ability (before the guns are turned against them) to pump tax revenues out of the captive public actually makes them a more popular client for international lenders than many of the democratically-chosen leaders who stride through their doors. By the time they leave office, much of the loot has usually been tied up in (hopefully) useless and un-resaleable armaments, guzzled on favour-repaying limousines and lunches, or spirited abroad into the exiled dictator's early retirement account. The inflation engendered by this excess often delivers a final blow by causing the currency to collapse, so that even the domestic revenue still being raised cannot buy enough hard cash to make payments on the debt.

Even governments which borrow modestly and spend wisely can be sunk beneath their debt loads by sudden shocks that transform the conditions in which they borrow. Once such crisis strikes, it becomes clear that borrowed funds are far from the string-free alternative to multinational-channelled inward investment which governments assumed when they (and state-backed companies) signed their credit agreements. Lenders may not dirty their hands with the day-to-day management of the finance they release, but their conditions for ensuring sound deployment and repayment can be just as draconian as those that big shareholders impose.

Most notorious is the International Monetary Fund's 'conditionality' – a standard set of actions required for emergency credit, drawn up with much closer reference to Chicago economic prescriptions than to local social and political conditions. These typically include deep cuts in public investment and welfare spending, sharp increase in domestic borrowing costs, reduction of corporate taxes and regulations, and withdrawal of subsidies from public corporations and farms regardless of the consequent job loss. The IMF is not intrinsically malicious, but its last place in the global credit system makes tough talk unavoidable. As the institution that countries turn to when all other credit sources have dried up, the IMF is uniquely placed to impose such 'adjustment' requirements on behalf of all creditors further up the chain.

If failure to meet these conditions leads to lenders pulling the plug, the distinction between debt and equity is effectively dissolved, as lenders exercise their right to grab real assets in compensation for unpaid loans. Creditor-appointed administrators push aside democratically-elected administrations to supervise the 'securitization' of overdue debt, by selling state assets to repay it or, if the assets prove unsaleable, by direct debt–equity substitution. Having made these swaps at rates which often register whole state-owned industries and commodity reserves as essentially worthless, governments can regain international financial market access only by following policies expressly designed to restore those assets' value.

Within months, if this succeeds, the creditors-turned-shareholders can find themselves selling their new holdings at many times the face value of the supposedly distressed debt. The buyers of these assets usually have little to complain about, however. They tend to be foreign multinationals who thereby capture operations that foreign governments had originally set up to rival them. For corporations previously unable to buy their way into key target markets because of impenetrable ownership structures, debt crisis can be the decisive chink in autarky's armour. Russia, South Korea, Cuba and Japan are among the most visible recent cases of corporate credit crisis stripping away anti-acquisition defences, and letting predators get their foot in the 'national champions" door. Anti-terrorist war is currently deterring, or drowning out, debt-driven protests in a number of crisis-hit countries. But their recurrence cannot be ruled out if over-indebted poor countries continue to be denied the time and extra funding for an orderly debt workout which over-borrowed rich-country companies continue to enjoy.

## The Costs of Letting Go

Economists have not honoured their original promise of enabling us to do things. But they are brilliant at explaining what isn't possible. Much theorizing and number-crunching was once put forward to explain how it wasn't possible to cure unemployment and inflation at the same time. Creating more jobs meant causing prices to rise, and stabilizing prices meant throwing people out of work. This gloomy outlook succumbed to a chink of light in the 1990s, when some economies managed to run themselves close to full employment without re-igniting inflation.

Unshaken, economists soon hit back with another, stronger dose of pessimism, arguing that a nation cannot reduce its unemployment and at the same time raise pay and productivity, unless it achieves an impossibly high growth rate. France seems recently to have raised both pay and jobs by shortening its work-hours, while China spent much of the 1990s achieving the impossibly high growth rate. But economists are unconcerned at this contrary evidence, because they have by now come up with the unsquareable circle to end all hopes of problem-free development.

The new message of doom is that once capital is free to move across borders, national governments can no longer direct their own economies. Before capital could move, it was possible to aim for 'internal balance' (full employment with low inflation) by regulating domestic activity through the interest rate. At the same time, 'external balance' between inflows and outflows on the country's current account could be achieved by adjusting the exchange rate. Once capital is free to flow, this becomes an impossible combination. If interest rates are pushed to a low enough level to promote sufficient investment for full employment, the exchange rate is liable to plummet, as foreign investors see their rate of return sink and pull their capital out. The effect becomes self-compounding, because the falling exchange rate raises the risk of inflation (further eroding the return on investment) and of difficulty for the country's government and companies in servicing any external debt. Investors respond to these threats by pulling more capital out, and the exchange rate keeps on falling.

East Asia in 1997–8 provided the most spectacular recent demonstration of this instability. The region's long-standing capital inflow became a fast-moving outflow as holders of its shares and bonds began to doubt the strength of its currencies, especially against the US dollar, in which most of them expected to be repaid. Those currencies duly dived against the dollar as foreign investors pulled back and local investors bailed out. Doubt had arisen because renewed slowdown in Japan had caused a slowdown in the region's exports, making it less clear that the capital goods that its firms were importing, largely with foreign loans, would generate enough foreign sales to pay for themselves. Companies had used the recent liberalization of capital markets to compensate for falling export revenues with new short-term loans. But now, that same liberalization allowed cold-footed creditors to pull their financial capital back. A similar pull-back by foreign shareholders pulled down the region's stock markets, setting off another round of defaults on domestic debt secured against shareholdings, land and other asset values. Because the region's economies were competing for the same set of export markets, the first to devalue forced others to do the same, to stay competitive. Under IMF pressure, central banks had raised their interest rates to forestall a self-compounding fall in the exchange rate. But by making it clear that their economies were heading for a profit-sapping recession, this simply gave foreign investors another reason to rush for the exit.

Making a country's currency cheaper, so that others buy more of its exports while locals cut their purchases of imports, was a popular way to reactivate growth in the days when capital stayed at home. Now, it is frequently counter-productive. For industrializing economies that are importing capital equipment and raw materials in order to export manufactured goods, the rise in the import bill when the currency devalues cancels out most or all of the rise in export revenue. Meanwhile, the weaker currency raises the cost of paying foreign lenders and shareholders back. And small downward adjustments in the currency sometimes explode into much larger ones, if they scare foreign capital into falling back to 'safe havens'.

If interest rates are kept high enough to keep exchange rates stable, and foreign capital flowing in, the result can be mass unemployment as investment and domestic activity slump. The domestic downturn will still, in the end, throw capital inflows into reverse, as investors see that any equity they hold will be in companies that can't make profit, and any loans they make will be to people who can't possibly repay. The exchange rate will still ultimately collapse, rekindling inflation and forcing default on foreign debt. Years of production growth and social progress may have been sacrificed in the meantime. Argentina in 2002 provides the most recent replay of this meltdown scenario. It abandoned its dollar-pegged exchange rate far too late to avoid a decade of recession under prohibitive borrowing costs, but amassed sufficient dollar debt under the false assurance of permanent parity to be rendered nationally insolvent when the unpegged peso finally fell to earth.

## Gone for Gold

The last prolonged phase of international capital mobility came to an end largely because of these instabilities. A recurring lesson is that, while floating exchange rates can be destabilized by shifting investor hopes and fears – amplified into a $3.5 billion-a-day foreign exchange market turnover – there is no safe way of permanently fixing the rates. European economies managed it early on in their industrialization, and imposed it on their various colonies, by retaining gold as an international currency. This aimed to remove one of the major risks of investing in another country – a sudden fall in the value of its currency, making the capital loss-making in home-country terms even if it made a positive return in the host country. But other risks remained: of the foreign capital being allocated to a use that found no market, destroyed in uninsurable disaster, or seized by local rebels against unpleasant colonial presence. Only when targeted at nearby economies with similar political arrangements and trusted social networks was investing abroad regarded as involving no more risk than investing at home.

But Europe's efforts to make its currencies convertible to gold again after the First World War, at pre-war parities, foundered in the downturn of the 1920s. The restored gold standard had helped to trigger this, countries having damaged their competitiveness (especially against the US) by fixing their rates, and hence national price levels, too high. After the Second World War, main trading nations tried again using a gold-exchange standard, with convertibility into the dollar. This allowed seriously misaligned nations to adjust their exchange rates instead of dropping out. But it still collapsed in 1971 when America was forced to suspend convertibility. By persistently buying more from abroad than it could sell there, the US had run up more payment obligations to other countries than its reserves could cover, if other countries asked for those dollar holdings to be turned back into gold.

Since then, Europe has returned to its single currency idea, with twelve states adopting the euro as of mid-2002. But others who could join the euro-zone don't want to. The monetary inflexibility imposed on disparate regions by the single currency is already forcing a relaxation of 'stability pact' restrictions on the size of budget deficit that member states can use to bolster growth when borrowing costs go too high. And those trying to stabilize their prices and attract inward investment by pegging to the euro are likely to experience – as have the various dollar-pegged nations – times when keeping the present parity forces up interest rates and scares off footloose capital, while adjusting the parity ignites inflation and derails debt-servicing plans.

Today, despite the recent retreat from exchange controls and reduction of capital taxes, capital must still cross most borders with much greater care. The abandonment of a gold-exchange standard, and proneness of the resultant floating exchange-rate system to unpredictable inflation and depreciation, has re-immersed foreign investment in substantial exchange risk. Investment risk also returned with the attainment of political autonomy by most former colonies. The first independence governments often took foreign-owned assets into public ownership, or used the threat of doing so to force changes in trading style that downed the private rate of profit. Their successors, though generally better-intentioned towards foreign investment, are still periodically forced to compromise capital's capabilities through austerity policies which stop it making the expected returns, or capital controls and currency depreciations which stop those returns from being brought back home.

## Uncorking the Bottle (Again)

Why, since the last world war, have industrial countries ventured back down the road of free capital mobility? And why, given their turbulent experience, have those still seeking to industrialize gone along with removing exchange controls? Given the patchy record of externally-financed growth, and the sacrifice of policy discretion needed to secure it, most nations might have been expected to target their interest and exchange rates and keep capital under wraps.

A short answer, from the anti-globalizers, is that new-industrializing countries were never given a free choice. Colonialism, administered by the European empire-builders, had locked them into the first global system of fixed exchange rates and free capital movement. Neo-colonialism, administered by American-led multilateral financial institutions, is charged with locking them into the second. But if the IMF and World Bank are on a mission to help multinationals flourish, their calculations have been notably bad. As well as losing much of its own investment when nations devalue or default, the IMF's response to the 1990s Russian and Asian crises was accused of worsening the reversal for private investors there. So annoyed was the Fund by the end of the decade at being the last-minute lender that helped private banks recover their debts, it began to insist on sharing any default-related losses with them. And when suspicion developed that its willingness to step in had encouraged reckless exposure by Western investors, the IMF sought to remove the 'moral hazard' by sitting on its hands the next time a currency crisis broke.

A longer and more likely explanation is not that the merits of capital mobility are still believed, but that those of discretion over interest and exchange rates have largely been dismissed. Government's power to manage demand for 'internal balance' is seen to have diminished because tax-and-spend powers have been trimmed by the electorate, and monetary policy powers by independent central banks. Its power to adjust the exchange rate for 'external balance' has been curtailed because a weaker currency, forcing wages and import prices upwards, hurts cost-competitiveness as much as it helps.

These ineffectivenesses of policy arise only because of constraints imposed by capital movement. But such movements are now generally seen as unstoppable, capital controls having been widely circumvented even before their formal abolition and now impossible to enforce in a world of electronic transfers and tax havens. This fatalism arises from two more common, and overplayed, beliefs about globalization: that it makes governments incapable, and that technology makes it inescapable. In practice, government is the missing ingredient for making capital flows work their magic. A firm political hand is needed to prepare the ground for foreign-financed operations, and to create the market conditions which ensure that they can sell what they produce. Globalization's gravest error is to think that it has outgrown the nation state.

## CHAPTER FOUR

# Making Too Much

*Economic critics charge the present globalization with under-supplying the social environment, while environmental critics accuse it of over-straining the natural environment. Since it's hard to consume too much while producing too little, combining both charges looks like critical overkill. But they do reveal two basic flaws in the way in which world production and consumption currently come together. Only by overcoming these can lower-income nations be enabled to catch up on income levels by exploiting their current wage gap, while richer nations find ways to live better while consuming less. Social forms of capital can stop the over-depletion of natural forms; but only if protesters see that what they want is not the disinvention of global capital, but the transition from a product- to a people-based version.*

Street parades are gaining pace, and losing their traditional politics. Big city centres used to be closed so that people in black, brown or red shirts could pass through them at a slow, deliberate walking pace. Now they are closed so that people in charity T-shirts can pass through them at various versions of running pace. As demonstrations speed up, they also change focus. Athletic contest has displaced political protest, as the goal that gets the masses on the move. The last century's serious marchers wanted a better life for their society. They would often

brave the pain of riot-police batons and overnight prison cells to get their message across. Their banners called for conflict resolution and reconstruction. Today's fun runners also want a better life, but primarily for themselves. They too endure pain, but in anticipation of a personal reward. Reconstruction is still an aim, but now an unashamedly personal one – stiff legs getting stronger, circumference fitting into ever slimmer shorts.

But runners are carrying on the street-fighting ideal, not betraying it. Real revolutionaries have always been impatient people. They believe that a system is fundamentally flawed, and doomed to collapse under its own weight of injustice or oppression. Yet they cannot wait for that collapse, and must instead try to bring it about. Lenin famously persuaded Russia's communists to 'give history a push'. Their disastrous direction of propulsion only increased the urgency and frequency of later rebels' chronological shoves.

Despite their conviction that corporate-led globalization is doomed, the present system's critics work night and day to bring about its downfall. If they really had the courage of their convictions, this massive effort would seem unnecessary. They could fly to Seattle, Genoa and Davos and see the tourist sights, not bothering with the conference sites – knowing that the plotting politicians would soon be out of office, and the conspiring corporations out of business, without any help. Instead, they act as if the runaway regime will cling to power unless protesters get together to engineer its fall.

This impatience is based on the sickening resilience of the present global system. The political, economic and cultural contradictions of capitalism have been known for centuries, yet it always survives – and responds to each wobble by reinvention on an even larger scale. Because Britain's Diggers and Levellers failed to sever capitalism's regional roots in the 17th century, it survived the Civil War to assert itself at national level. Because Russia's Bolsheviks couldn't bring down its national incarnations at the start of the 20th, it had raised itself to continental level by mid-century. So unless the new economic integrations of Western Europe, North America and East Asia can be stopped in their tracks, they will forge a global order during the course of the 21st.

To its supporters, linking national economies makes a good system better, and closer political ties make sense as economic growth gives nations more shared problems to work on together. To stop the globalization process now would, as well as locking less prosperous countries into their economic disadvantage, leave the whole world less well off than it could be. But to opponents, globalization just postpones free market forces' day of reckoning. The fundamental problem will soon be back, they warn, made bigger by postponement. Reviving an older line of reasoning about economy and society built on self-interest, they see a basic imbalance. The system must run to stand still, but can ultimately run only into the ground. In case one explanation for this is not enough, the protesters offer two: under-consumption and over-production.

## The Spectre of Under-consumption

The economic explanation, developed by Karl Marx in the 19th century and repackaged by John Maynard Keynes in the 20th, is that free-market societies have a natural tendency to underuse their human resources. At most times, many people will be left out of work, because the system cannot make it profitable to employ them, despite the many useful things they could do. At all times, most people in work are producing far less than they could, having been forced into jobs they don't like, making things few people want, and are unable or unwilling to do more than the minimum for survival.

So capitalism condemns itself to make more than it can consume, because people are unable or unwilling to spend all of the income they generate. Marxists blame the under-consumption on an *inability* to spend, and trace it to inequality in the current situation. Employers make their profit by holding back a part of the workers' product, the 'surplus value', paying them less than is needed to buy all of the output they make. The gulf gets progressively wider because industry keeps increasing the rate of surplus value, by making people work longer and harder at ever more simplified and speeded-up tasks. Even when, to stop a rebellion, companies start paying their staff more than is necessary to keep them alive and working, the things they'd like to spend on are often unavailable, because meeting basic needs doesn't give product designers and marketers as big a reward as catering to luxury tastes.

Keynesians blame the demand deficiency on an *unwillingness* to spend, and trace it to lack of clarity in the future situation. Doubt about what their earnings and preferences are likely to be leads people to keep money aside, for times when they might need to spend more than normal, or more than they receive. Even if there are rewards for investing spare money, drawing interest or dividends from the person who uses it, the desire for 'liquidity' leads people to hoard it as idle cash.

Whether through Marx's 'cash nexus' (human relations' reduction to one-sided financial exchange) or Keynes's liquidity preference, the result is a recurring problem of too little buying power chasing too many goods. Present unfairness or future uncertainty stop people from buying all the fruits of their labour. Companies can't sell all their output – so making the profits that repay past investments and underwrite future ones – unless something closes the gap. A source of consumption without production is needed to fill the void caused by production without consumption.

Trade-based societies have evolved a number of mechanisms to tackle this problem. These involve regular 'injections' to offset the 'leakages' of aggregate demand, whether these arise through exploitation or precautionary saving. When agriculture still dominated the economy, the solution lay with landowning barons, or Lords of the Manor. Their onerous task was to collect, as rent, any money or produce that the peasants had left over after meeting their subsistence needs; and to spend it on banquets, balls, jewellery, mistresses and other luxuries, while staunchly resisting the urge to do any productive work. Unfortunately for the stability of this system, some of the landowners got enterprising with the rent they collected, and re-invested it in mines and machines which started adding to the flow of goods. Others, by failing to spend enough on the heavies needed to collect the rent, or to index those rents to inflation, left too much spare cash in the hands of the peasants, who were also inclined to use it to diversify their production instead of multiplying their consumption.

These oversights eventually triggered industrialization, as landowners and merchants saw factories as a better place than fields to plough their surpluses into. Peasants forced off the land by the new mechanized intensive farming flocked to the cities, and windowless workplaces where hours were often longer and far more intense than any they'd previously endured in the open air. This greatly raised the new industrial workers' productivity, and so amplified the Marxian and Keynesian demand gaps. With much wealth now at stake, new tactics were quickly developed to fill them.

An early route was through imperialism. Annexation of other land in search of gold or gods pre-dated industrialization, but was easily turned to its advantage. By pursuing the newly-discovered territories' rich ruling élites, and persuading them to spend abroad the wealth they siphoned off at home, colonizers opened up another source of demand for industrial goods that wasn't inconveniently linked to either present or future supply.

A less unjust, and more sustainable, approach was to finance investment through borrowing. New buildings and machines could be built with money that didn't yet exist, with a promise to repay later, from whatever the project earned. This filled the gap between present production and present spending by bringing forward some of the spending linked to future production. But it only pushes the problem one stage back. Investment spending (by private enterprise or the government) initially puts more money in people's pockets without putting any more goods on the shelves. But if targeted wisely, it will eventually lead to more production. After demand's initial boost, supply runs ahead of it again. Worse, the investment tends to be in new machines and management techniques that wring more output from the people who work with them. So each short-term boost in demand achieved by this method is followed by disproportionate rise in longer-term supply.

This problem is avoided if new investment is rendered 'unproductive' – channelled into capital goods that no one wants to use, or whose output no one wants to buy. Making existing capital stock redundant can also help to get investment's demand boost without its capacity expansion. This has most commonly been done technically, by inventing new products and processes that out-perform the old ones; or pyrotechnically, by destroying the old ones in acts of war. One historically recurrent device, the arms race, neatly achieves both of these effects. Entering an arms race allows nations to channel their surplus capital into weapons of mass destruction, which add nothing to productive capacity. Periodically unleashing those weapons on each other brings a massive mutual reduction in productive capacity. The stage is then set for rebuilding the factories, restocking the arsenals, and building up living standards from a much diminished base. So once the financial mess is sorted out, investment can recover until employers' production power again outpaces exploited employees' consumption power.

War has side-effects, however, which makes this an expensive way to tackle over-production. Even with precision-guided missiles, war destroys far more than surplus capital. People, and plants of the natural kind, tend also to suffer, and aren't as easily replaced as mangled machines. War breaks supply chains and disrupts trade flows, by arousing political emotions which make it hard to achieve the obvious new profit-making opportunity of selling weapons to both sides. Although armed conflict remains a common way of correcting the consumer society's overkill, less painful means of investing without producing have recently been found.

One involves swapping weapons of mass destruction for 'gales of creative destruction', the much-quoted characterization by Austrian economist Joseph Schumpeter of capitalism's tendency to cull its old ways when something better comes along. Innovation is inherently wasteful, most new ideas closing down more technological avenues than they open up. Investors in risky new ventures have long acknowledged that fewer than 10% of them ever pay off. In the past, they tried to get more for their money by better spotting and screening out of the loss-inflicting 90%. But in the 1990s new gadgets started arriving so quickly, and the public became so fickle in its choice of which to accept, that the winner-picking approach gave way to the scatter-gun system of product development: treating all innovations as if destined for a 'killer application', and leaving buyers to decide which they actually had a taste for. The no-hope projects were showered with capital as liberally as the fail-safe, with the unviable and the enviable mixing freely in marketers' minds.

Internet commerce proved the first favourite dumping-ground for wasteful capitals. For several pre-millennial years, the new economy destroyed far more value on 'dot-coms', through far more outlandish or pointless escapades, than the military could ever have wasted on its smart bombs. Initial excitement over business-to-business (B2B) and business-to-consumer (B2C) uses of the Internet spread to other untested techs over whose fanciful accounts hard-headed financiers were prepared to suspend their disbelief. Genomics, proteomics, plastic and optical electronics, fuel cells, virtual reality, third-generation mobiles, application service providers and Internet shopping all opened up as bottomless pits into which the old economy's surplus savings could queue to despatch themselves. Unwary share buyers didn't need to dirty their hands on the technical details. To perform capital's system-saving vanishing trick, they merely sank small fortunes into fast-rising stocks when dot-com delusion was at its height, and waited for a miniaturized fortune to emerge when the reality of redundancy sank in.

Japan, disqualified from the arms race by its mid-century adventures, had already demonstrated in the 1980s how to lose spare cash in a flash, by pouring it into office and factory buildings whose prices then plunged through their own empty floors. America achieved an almost equivalent asset deflation in the late 1990s without even having to buy the land or mix the cement. Registering an unpronounceable name and designing an unnavigable website were enough to do the vanishing trick, until the corporate red ink could be wrung from the cyberspatial black hole.

Where dot-coms did their bit for the capital cull, television made a follow-up contribution, extending the value-destruction venue from Silicon Valley garage to suburban front room. The entertainment industry, always a monument to consumption without production, tends to have its best times when other sectors are shedding labour, leaving plenty of people to shelter behind headphones or slump in front of screens. But with Cold War seriousness expunged from public debate, many other televisual products (including politics, drama and journalism) have been claimed or reclaimed by the world of entertainment. And many real-world activities have come to rely on an entertainment dimension, from computers that come bundled with game software to gambling based on stock-price movements.

Entertainment products, like weapons systems, absorb resources during production without yielding a stream of goods or services after completion. The most that a programme or painting can produce are ephemeral sensations in the eyes and ears of viewers, just as the most that can issue from an arms dump or missile silo is the intangible assurance of a life-saving balance of terror. By diverting creative effort towards mass-produced films, wall-to-wall pop warble and a thousand cable channels, much of the horror and numbness of war can be reproduced without leaving home. People pay for the TV experience, yet come away with nothing to show for it except square eyes, jangled ears and lust for holography's latest heart-throb. After dot-coms, sit-coms have become a valuable forum for value destruction, and the technique is now taking off elsewhere in the art world. Gallery-goers pay to inspect 'installations' indistinguishable (if not actually made) from junk lying in the street that they could see for free; 'reality TV' fans queue to view their own lives scripted and censored (for a fee) as soap opera; adding one glass bottle and the name of a distant hill turns unremarkable water into a potion worth two thousand times as much as what comes out of the tap. We can't believe what's happening until we see it on film. And in the process of getting it there, we help the economy to correct its consumption gap by paying for more of what had never needed saying before.

## The Ogre of Over-production

Just in case these economic accounts of capitalism's downfall don't convince, or don't make sense, its opponents can now offer an alternative, ecological, explanation. This has the advantage of appearing, at the same time, both the mirror image and the polar opposite of the economic account. If one condemnation is dismissed, it's therefore hard to stop the other being embraced. Solutions offered by the ecological case can, similarly, either dispose of dissident economists' solutions or dovetail with them. The way out of the mess is either to turn back to pre-industrial lifestyles, or fast-forward to a post-industrialism in which slashing and burning are no longer the disorder of the day.

The environmental explanation for free markets' eventual economic failure is their social tendency to over-use natural resources. Pioneered by clergyman Thomas Malthus as the 19th century opened, it was revived by less religious but equally evangelistic 'green' campaigners at the close of the 20th. For the 'greens', the gap that exists between the available and the affordable is essentially the same as that shown by Marxian and Keynesian economists. But the 'green' approach seeks to close the gap the other way. In this view, citizens cannot consume all that industries produce because production runs ahead of what the planet can afford. If we don't slow down, making more with less or making do with less, an end to natural growth – as topsoils dry, forests die, seas coagulate and air pollution twists the global thermostat – will put an end to economic growth.

The essential problem is that markets, a device for allocating infinite flows, have been brought to bear on strictly limited stocks. In under-pricing exhaustible natural resources, by failing to acknowledge their scarcity, 'post-industrial nations' are consuming their seedcorn while believing that they're having a feast. When too few purchasers chase too many goods, it is the volume of goods that must be managed downwards. (One deeper shade of greenery also questions the planetary carrying capacity – daring to suggest that while the whole world population might still be able to stand on the Isle of Wight, there's now so tight a squeeze that the Japanese would start complaining. But any suggestion that our planetary conservation quest should extend to controlling our own numbers gets far closer to Malthus' passion-killing mission than most neo-Malthusians are now prepared to go.)

In the physical world, economists distinguish created means of production (industrial capital) from the flow of goods resulting from that capital, and count output as 'sustainable' only when it leaves capital intact, so that workers can breathe again and production lines roll another time. But in the natural world, according to the green approach, they have missed the similar distinction between inherited means of production (natural capital) and the flow of goods resulting from that capital. Running down the stock is thus mistakenly viewed as a way of speeding up the flow – until those who return to repeat the process suddenly find that there is no more oil to burn, no more crops to harvest and no more fish to fry; or that practitioners choke as they do so, because equally scarce clean air and water stocks have likewise been run down.

Green movements first sounded the alarm over binding shortage of irreplaceable inputs to production. Excessive output would be cut down to sustainable size because there would be nothing left to feed the factories, or fuel the distribution chains. In a world pulled up short by a politically-created scarcity – the first OPEC oil embargo – a distinguished forecasting group commissioned by the round-table Club of Rome famously proclaimed the dawn of naturally imposed scarcity. Affordable supplies of petroleum, whale-meat and several widely-used metals were all supposed to run short by the early 1990s, according to the Club's cold-blooded computer calculations. Rome's neo-Malthusian revenge was embarrassed by certain omissions from its model. Insufficient acknowledgement was given to how far the rising price of exhaustible commodities, set in train by approaching scarcity, would spur extra effort to locate and extract more supplies, and to develop more abundant substitutes. By the late 1990s fossil fuel was still flowing, food production was still rising, and the economic optimist Julian Simon had won his bet with the ecological pessimist Paul Ehrlich about supplies of key natural commodities going up rather than down.

More recently, environmentalists have found themselves spoilt for choice over scarcity, with many a non-material item also threatening not to go on *ad infinitum*. The problem now is not nature's cupboard going bare, but its rubbish sack overflowing, because of a surplus that Marx hadn't noticed – of unintended outputs from production. Just as it under-prices, and so over-uses, nature's below-ground bounty, industrial production puts insufficient value on the power of air and water to absorb its waste. So toxic landfills overflow, carbon sinks overspill, the ozone layer thins and the city air thickens. Well before oil and gas run dry, we are warned, the by-products from burning them will have warmed the air and parched the soil enough to undermine the way of life that runs on them.

That way of life, environmentalists add, is equally doomed whether its industries roll beneath a capitalist, socialist or intermediate 'third way' banner. The communism that set out to displace market-based Anglo-American arrangements, and developmental states that tried to tread between the excesses of the two, turned out to be no better at fairly pricing natural products and stemming the ruinous relegation of stocks into flows. Industrialism and internationalism are the twin enemies of sustainability, whatever their ideological guise. So whereas underconsumption stories take the left-leaning view that all would be well if class barriers were transcended, over-production accounts tilt rightwards, with planetary physical limits assumed to kick in whichever side of the spectrum claims to be in power.

## Erasing the Standard

Any 'race to the bottom' is an uncomfortable contest. All too often, the winners end up envying the losers, seeing prize money drop when the national league gives way to a world championship. Anti-globalizers predict disaster when high-paid labour is thrown into direct competition with developing-world equivalents on a tenth of the wage. But once trade barriers fall, participation seems compulsory, even for those with illness or handicaps needing treatment from the sidelines. Once-insulated bankers and builders are thrown up against flown-in substitutes, or their practitioners; surgeons and receptionists must confront low-wage counterparts who ply their trade down the phone. The rich world's natural parks come under pressure from foreign industrial parks, where less highly valued greenery has been ploughed up to make cheaper production sites, or logged to cut the cost of raw material. First prize, a week's vacation in a *maquiladora* on the Mexican border. Second prize, two weeks' vacation.

But those who take the shortest cut, whether through devalued employment or a depleted environment, have not been noticeably more successful in this race. If lowering labour costs or environmental standards were the key to global competitiveness, capital would not flow so readily from low-income areas to high. Conversely, development is a race in which rich sponsors count. Higher wages, on a world scale, tend to be backed by higher skills, which support the extra pay with extra productivity. A wealthier workforce can also, despite its electoral turn towards the tax-cutters, support a stronger public revenue base. This enables government to offer employers productivity-enhancing transport, communications, healthcare, education, policing and public planning systems, to standards unavailable in most low-wage states that they might be thinking of migrating to. The admission of Mexico to Nafta, and of Eastern Europe to the EU's single market, did not spark an exodus of footloose capital or a bidding down of wages as existing workers fought to stay in demand. Low-skill labour in the high-wage area may have suffered wage and job cuts, but their counterparts in the low-wage area experienced a similar upward pull on pay.

The charge of 'dumping' artificially cheap products, to disrupt someone else's market, is often over-used. An offer that undercuts local competitors through being delivered at lower cost is just competitively priced. It is only 'dumped' if price falls below cost, implying government subsidy or corporate cross-subsidy. Even then, conventional economics says that we should welcome dumped items, not raise tariff walls against them. Someone else is effectively paying us to take their goods away, so we should do so. 'Anti-dumping' protection is only socially justified if the cheap goods threaten an industry or community that cannot easily regenerate. It is only economically justified if the subsidy is part of a foreign firm's monopoly price discrimination, or designed to knock out local competition so that prices can later be raised to monopoly height.

'Social dumping' accusations take the charge-sheet much wider, however, because as well as prices below costs, opponents can cite costs that fall below a common decency threshold. If a foreign company can ask for less because it gives its workforce less, efforts to compete with it could spark a tit-for-tat wage-cutting battle which leaves everyone worse off. Better to prevent this, the dump-detectors argue, by forcing foreign companies to pay a higher minimum wage or improve their workers' non-wage treatment. Or if this cannot be done, to raise the tariff on imports until they cost the same as if those making them were paid respectably. Inevitably, the wage-raising case is usually put by leaders of well-off trade unions in rich countries, who want to preserve their members' present shelter. Cleverly, it is presented as being in the interests of the low-paid workers behind the 'dumping', whose leaders can often be summoned to endorse the all-round higher-pay claim. The favour is easily returned: when British coal-miners or American car workers demand a big rise for themselves, they have nothing to lose by demanding a similar jump for Polish or Brazilian counterparts – and possibly a job to lose if their own increase isn't matched this way.

Instances of MNCs outsourcing products to cheaper foreign locations, or using this threat to force down domestic wages and prices, are not hard to find – nor, as the anti-globalism campaigners have been finding, hard to publicize. Public outcry is especially easily aroused when those to whom the jobs migrate are under-age as well as under-paid. But shifting work to cheaper labour does not have to mean exploiting it. Both sides can gain, if those receiving the jobs can push up their income in line with their rising efficiency, while those conceding the jobs manage to move to higher-productivity use. Arresting the process by hiking poorer countries' wages will not help them to escape the poverty trap, unless their employers get more positive help towards matching gains in productivity. Offsetting the process by raising tariffs on cheap-labour goods swells the rich importer's tax receipts while cutting the poor exporter's take-home pay. Children are especially at risk from over-zealous anti-dumping action. Many have found themselves redeployed to even worse workplaces, rather than returned to their rightful home or school, when the bar goes up against products of child labour. Putting an end to this requires positive help to subsidize schooling and raise adult workers' productivity, not just the negative refusal to engage in pre-teenage trade.

Most relocations of production have been part of a long-running process in which labour-intensive industries go where labour costs are lower, leaving higher-paid labour to seek higher-productivity use, often administering the new overseas operations. Where sweatshops have been used, the target market has tended to detect and object. Users' refusal to pay because quality had followed cost downhill, as much as unions' objection to the new suppliers' prison-camp conditions, were the reason why Nike had to raise its Mexican sub-contractors' standards, and Marks & Spencer reinforce the rule-book at its South-East Asian plants. If any race has developed, it is towards the middle or upper level, with newcomers to the international market having to train more and pay more to stop their better staff from leaving. Without such positive counter-actions, efforts to block low-cost imports – while sometimes a genuine attempt to stop global driving-down of pay and conditions – are often just another disguise for the privileged world keeping its protections up so that low-income counterparts can't build on their greatest competitive strength.

The only way out of the race is to recognize these relocations as mutually constructive, difference-narrowing adjustments, and swap head-on competition for pursuit of complementary tasks. This is exactly the way that these free-trade groupings have gone, once their richer half persuades itself that jobs lost to cheaper parts will be replaced by something better; and that if the skill-gap-closing challengers aren't allowed the work that's due to them, they'll come across to get it.

Steel-makers cannot turn quickly or costlessly into software engineers, or car assemblers retrain as consultants overnight. Living with the new international division of labour does not, as free-trade futurologists once hoped, simply mean production workers switching to service jobs at the same levels of authority and skill. Those made redundant when a major plant moves abroad cannot make the same living delivering pizzas or driving taxis for each other. And even if shopping malls or call centres can eventually employ the same numbers at comparable salary, these service jobs are proving equally transplantable to places where telephone voices are cheap. The only lasting escape, once nations emerge from behind the barriers that blocked wage and price convergence, is for those whom the unified market finds no longer affordable to climb higher up the value chain.

Present wealth, and possession of the less exportable forms of capital, gives the rich world the resources to manage this transition. To do so, however, it must absorb two major social costs. Early retirement support is needed for an older generation which cannot adapt to the new workplace. Better education, training, health and mobility must be promoted for the next generation, so that it can slot into its more demanding roles. What high-income nations currently lack is not the resources to fund this social upgrading, but the resourcefulness to plough their current wealth back into it. Those now in comfortable late career have been taking – in tax cuts for their own consumption – precisely the resources that those just starting out need publicly invested to keep themselves competitive. Conflict between old and new work generations at home is potentially much more serious, and long-running, than that between old and new industrial nations abroad. The world's leading economies should win the battle within themselves, over how to prepare collectively for the reconfigured work world, before they contemplate trade war against emerging economies that are only trying to catch up.

## Down in the Dumps

Better-off regions put a higher price on their environment, as well as on themselves. Those who want tariff protection extended to 'environmental dumping' point to labour costs brought down by lack of concern for what staff handle or inhale, and material costs minimized by harvesting and mining at unsustainable rates. By pricing their environment low, developing countries can concentrate production with the hungriest production lines and dirtiest smoke-stacks. If they won't put higher value on irreplaceable materials and carbon sinks, richer trading partners reserve the right to do so, to stop a general under-pricing of scarce resources that could leave the whole planet gasping for breath. But doing so with tariffs again brings the danger of discriminatory, inequality-enshrining protectionism under a more green and pleasant name.

The only 'sustainable' way to spread best-practice environmental standards is, as with labour standards, to level up living and working conditions. Industrial nations have started to see that this is best done through unilateral action to price environmental inputs sustainably, even if at first this seems to widen the advantage of those who slash-and-burn. The EU's energy taxes and recycling targets reflect the realization that an early encounter with the higher oil, gas, plastic, paper and parking charges of the future will give more time for adaptation to conservation in these areas. Japan ran an early advert for this strategy, its curbs on auto exhaust emissions dating from the 1970s helping to underpin its global lead in small-car exports by the early 90s. As lead-free petrol and CFC-free packaging shows, industry can move quickly to replace poisons once they're priced out of the market. But it rarely does so until comprehensive green taxes and clean-air regulations give it no cheap-and-nasty alternatives to turn to.

Environmental innovators face the usual first-comer problems. Others can copy their new technology without the development cost; and the first offering from that technology will be too expensive to compete with current, more polluting products. There is room for social sharing of these costs, through subsidy to the development and initial provision of greener alternatives. The present scale of these subsidies is tiny compared with those that go to existing, environmentally backward-looking technologies:

renewable energy still gets a fraction of the subsidy received as tax breaks for conventional energy investment, and the whole of a city could be made fit for bicycles with the budget required to build another underpass for cars. But markets already give incentives, at least as effectively as any other allocation arrangement, to economize on resources whose supplies are starting to tighten. Major lifestyle shifts – from wood to fossil fuel when forests shrank, oral to audiovisual culture when attention spans narrowed, bus to car when journey-time ran short – have been propelled by little else.

Richer nations long ago shifted from 'extensive' growth based on mobilizing more resources to 'intensive' growth which gets more out of those resources. Denying nations' inability to think across in space or ahead in time, the 'green growth' counter-lobbyists point out that major coal- and oil-burning nations signed the Kyoto protocol on greenhouse gas emission cuts, and some may keep their commitment to its targets despite the biggest polluter, the US, refusing to ratify. Contesting the link between producing more goods and consuming more natural resources, 'green growth' supporters point to examples of the link becoming inverse. China's growth through the 1990s may not always have reached the double digits shown in its statistics, but its economy and population undoubtedly expanded, yet its carbon dioxide emissions had fallen by the end of the decade. Having learnt to live with high wages by lifting labour productivity, the world's rich must now accommodate raw material scarcity with a similar rise in resource productivity. This means putting a higher price on their own scarce natural inputs, not just trying to raise their price elsewhere.

## A Non-Material Future

'Green' growth is helped by changes now occurring in the form of goods and capital, which get less tangible as nations grow materially richer. Today's enterprise trades on, and holds together by, what it *knows* as much as what it owns. Commercial initiative passed in the 1990s to 'virtual' companies, networks of alliances whose fixed assets extended little further than a small head office and a suite of small computers. Cisco encircled the world with cable that it never had to buy and sell; the Rolling Stones became a global transportation and delivery service just for the duration of their world tours; the Virgin group's lightning strikes on possible markets, from bridal wear to vodka, were enabled by the speed with which suppliers and distributors could be hired in and just as speedily paid off.

Virtual firms bought inputs from outside suppliers, passed them to sub-contractors for processing, and sold the results through franchised retailers, minimizing the risks of 'working' capital investment by ordering on sale-or-return, and of 'fixed' capital investment by leaving others to make it. In the 2000s, these 'supply chain integrators' changed their strategy again: they sold and leased back the small head office and the suite of small computers. If you're ever bought lunch by a virtual company director, note how she invariably wipes her plate clean. An unmunched broccoli clump could wreak havoc with her diligently voided balance sheet.

Capital has changed, from commodity to concept, from industrial plant set in concrete to a post-industrial abstraction. Its role remains the same: a stock of human-created devices which help to deliver goods and services that people will pay for, without being used up in the process. Its form has moved on. Not because technical advance made the old ways coarse and clunky, but because capitalists feel safer holding wealth in the privacy of their heads than in the palm of their hands.

The 'de-materialization' of capital has long been attempted by people who saw that wealth held in physical form could too easily be wrested from them by the four awesome forces of robbery, destruction, obsolescence and progressive taxation. Their first defence was the conversion of physical capital into 'finance capital': money, and various paper assets almost as accessible as money because they were available in small units and could quickly and cheaply be bought and sold. Foremost among these assets were corporate shares and bonds (giving entitlement to profit and interest earned by private companies), government bonds ('earning' a return ultimately backed by the taxpayer) and bank deposits (rewarded by the interest charged to the people and companies that borrowed from the bank). Financial assets were, in most circumstances, economically safer than physical ones because they were easier to hold as part of a portfolio (spreading risks which were otherwise concentrated if the same amount were tied to owning one company), and easier to re-sell. Shares also give entitlement to income from the whole of a company, which must be much more than the sum of its physical parts, or else raiders would already have descended to merge it with others or break it apart.

The financial form of capital is still vulnerable to robbery (by thieving individuals or nationalizing governments), destruction (by inflation or company collapse which renders it worthless), obsolescence (when withdrawn too late from investments that have lost their market) and taxation (financial stocks and flows being generally easier to identify and tax than those in physical form). This is not the way that most successful citizens wish to hold their wealth. The logical next defence is to re-invest in themselves, using their financial wealth to acquire human, cultural and social capital. Knowledge and networking are the post-industrial equivalent of gold and silver jewellery, the easiest way to keep the new capital safe and carry it around.

This would take the owners of wealth full circle, to a form of capital whose commercial risk – like the machinery with which they started – cannot be separated from them, so that they cannot be separated from the capital's commercial application. The key difference is that, for the non-material forms of capital, this application is more to consumption than production, more pleasurable than painful. Human capital can build up in the classroom with little of the chemical and social stress that accompanied physical capital accumulation. Cultural capital can generate its 'income' flow from little more than a fume-free painting, book or screen. De-materialized capital opens a route to more environment-friendly production through a shift to environment-friendly investment and consumption. Going down this route means raising people to a level of wealth at which materials' form is as important as their function. This requires an accelerated passage down industrialization's growth path, not restrictive action to stop its worldwide spread.

## Double Jeopardy

Accusing a culprit of two opposite crimes should make it easier to secure a conviction on one of them. Globalization's effort to beat the under-consumptionists' red light, by expanding present demand (and periodically culling excess capacity), means a heightened risk of hitting a speed trap at the over-productionists' green light. Because of its refusal to pay out what its human resources are worth, or work out what its planetary resources are worth, the present world system is damned on demand and equally damned on supply.

Globalization's arithmetic doesn't add up. From this flows the hope of final triumph for those who seek to trash the Nike flash. When outsourced shoes cost five cents to make and 50 dollars to buy, there's an economic imbalance that undermines the global production system. When nations with the least capital must send it, as investment or interest payment, to those who already have the most, there's a social imbalance which drives action that could hasten that arrangement's collapse.

But where standards of proof require consistency, the 'red' and 'green' witnesses claim too much. If all production leads eventually to consumption, the global economy cannot be charged both with under-consumption and over-production. Another U-shaped curve, with rewards for those who suffer through the mid-term crisis, is preached by pro-capitalists in relation to growth and the social and natural environment. If other ways are found to let people buy what already exists, they won't have to rely on a system that must currently prepare to produce ever more.

The argument that the system will get healthier, if allowed to grow a little more, rests on two problematic requirements. The first is that for growth to be green, new investment must make capital, as well as labour, more productive. Recent trends – especially in the US – for productivity to grow despite shrinking investment suggests that innovation is saving on capital as well as labour. While good for the ecological balance, this presents further problems for the economic balance between income and expenditure, which depends on accumulating more capital to pay for the products of existing capital. If preparing for tomorrow's production costs less today than it did yesterday, current production could run even further ahead of current consumption. Moore's Law, the perennial doubling in power and halving in price of microprocessors which underlies the steady price fall of products that use them, risks resurrecting Marx's law of oversupply and falling profits. To get people to the level of wealth where they can afford to be green, private activity must be better coordinated and strategically supplemented. This points to a rediscovered role for the state, to be assessed in chapter five.

The second weakness concerns oil. For all their talk of a post-industrial transition, high-income economies are burning more of the black gold than ever before. The progenitors of corporate-led globalization – major oil, mining and chemical companies – are back to their predatory worst, delving even into rainforests and ice-caps to keep up the flow of cheap petroleum and gas. Political strategists aligned with them are following a similarly reminiscent plan, using the war on global terrorism and reconstruction of 'failed states' to justify annexation and occupation of lands which happen to lie above promising reserves or in the path of needed pipelines. But as the ructions over Kyoto and Kabul showed, this is a peculiarly American problem, and one that will be explored more fully in chapter seven.

## CHAPTER FIVE

# States of Disarray

*Globalization leaves governments restricted in their scope to tax, spend and regulate, but only because they wish to be so. Pretending that it's possible to pass old public duties to the private sector, or impossible to do them because of outside-world constraints, is more comfortable than tracing tough choices to voters' own contradictory desire to get more and pay less. But while today's rudimentary international governance institutions are easy targets for the anti-globalists, they are the best available route to the global regulations and global public goods provision that must now be developed. WTO, G8, even IMF are not grim acronyms deserving destruction, but one possible route to recreating at global level the necessary stabilizing role that national governments never lost.*

To critics of the borderless business world, the counterpart to rampaging multinationals is a retreating national state. Political action demands national boundaries that private enterprise breaks down. Leviathan needs downsizing if Mammon is to grow. It is in anger at spectating at this zero-sum game, between democracy and plutocracy, that anti-globalizers stage their stadium riots.

But the present meshing of national economies is happening with governments' consent, not against their will. Globalization is a political project, transcending but not ending national sovereignty. Politicians' proven self-preservation urge ensures that any powers lost to presidents and parliaments will return with interest at supra-national level. At present, like most activities *within* today's national states, relations *between* states are being subjected to a privatization process. Responsibility for cleaning up international relations has been passed to Procter and Gamble, diplomatic bridge-building spun out to Balfour Beatty. Since globalization heightens economic and social imbalances that call for more public action rather than less, the task of rebuilding and re-legitimizing the state is now becoming urgent. There is now a chance of recreating, at international level, past government successes at promoting national development. But only if protesters stop disrupting the talks at which this new system's architecture is being hammered out.

## Unaffordable Government

Many traditional supporters of the strong national state believe that there's no longer much it can do. Its critics add that, of the things that it can still do, most are better left undone. Far from being something that existed from the dawn of time, today's state now looks like a fleeting creation that's outstayed its welcome. When more people trust their local supermarket than their member of parliament, the eclipse of cabinet government by corporate governance begins to look preferable as well as real.

In the 'can't do' category, governments are losing the ability to raise revenue to finance the expenditures that they used to take on. It was always a struggle. When national wealth lay in land, its owners routinely resisted all efforts to tax it. When they finally paid up, it was only after first creating a parliament in which the then monarch's use of the money could be checked line by line. But land is, at least, hard to conceal, and even harder to move. So sovereigns could see it and slap a final demand on it. Now that wealth derives from labour and capital, those monarchs' successors have a king-sized hole in their pocket.

Businesses can shift their operations so that profit is highest where tax is lowest, or use 'transfer pricing' (routing the corporate treasury through a tax haven) to give the appearance of doing so. Most have two sets of accountants, one showing the shareholders how well they're doing, another to show trade unionists and the taxman how hard the times are. Taxable profit evaporates at the end of the tax year faster than a snowman at the start of spring. Because they have to carry passports, and like to put down local roots, people have less scope to move around in search of tax breaks, and less hope of finding accountants who can tap them all. But the scope for governments to claw money back through extra personal tax demands is similarly limited. If they can't avoid high taxes, people can still cast votes against them, and a kick in the ballot box is as painful to most leaders as a corporation voting with its feet.

Governments traditionally borrowed to keep up their budgets when the tax didn't arrive. Much of the money that the world still trades with started out as debt secured against the national treasury. But as governments show that they can misspend the money and dishonour their debts as brazenly as any private borrower, they've had their loan costs raised and their overdraft limits cut.

Reluctance to have their hard-earned money extracted through taxes reflects people's growing belief that they could spend it better themselves. A once impressive range of 'public' goods, things that the government had to provide because no one else could, has been keenly pared down as non-state alternatives spring off the drawing-board. From an economic viewpoint, education should be financeable through private loans, because it gives people skills that can raise their later earning power. Public transport ought to cover its own costs once private motoring comes within majority reach. Housing and healthcare, which may have needed subsidy when many were still living close to subsistence, can safely be sold at market price once growth has given everyone the ability to pay. Even defence and policing, once collectively funded because everyone gets the benefits, can become fee-based once more accurate missiles, and crime's migration online, allow protection to be limited to those who are up-to-date on their subscription.

## De-Nationalizing Decision

Even where there's still an economic case for public funding of some services, public opinion has swung against public provision. It's not that politicians and bureaucrats aren't trusted to chase value-for-money as assiduously as private managers. Rather the reverse. They've been proven only too good at lining their own pockets from the public purse. State officials' public image as selfless servants of the social good was long ago lost amid rumours of leaking Treasury coffers. Even the incorruptible Germans have seen top politicians felled by corruption scandals, while America's current presidency sometimes seems like one long round of repaying the sponsors. It's a self-fulfilling suspicion: the less people trust their public servants, the less well they're paid, and the more they'll need bribes and backhanders to take home a living wage.

Even where they're not corrupt, state decision-makers face an uphill struggle to look efficient, because of constraints in public choice less often suffered by the private sector version. Unlike business leaders, politicians can't easily prioritize one issue. An embattled multinational can vote to downsize its Mediterranean operations, approve a 5% pay deal with line staff, set a launch date for the autumn range and select beige and orange for the new canteen, while the government is still horse-trading over discussion time between the annual review of salmon subsidies and the emergency debate on troop deployments in Sierra Leone.

When the debate finally kicks off, boardrooms with their strict shareholder focus can reach key decisions in a tenth of the time it takes most cabinet rooms with their cross-cutting constituencies of backbenchers, voters, lobby groups, advisers, foreign governments and opposition counterparts with favours to return. Meetings, the stuff of political life, are a distraction from the business life, so private practitioners make sure they come quickly to the point and even quicker to the vote. Without, if possible, the blanket publicity campaign needed to persuade people it's no worse than the system it replaces, or the fifteen amendments required to make it work.

In consequence, top management teams can often roll out a new product in a fraction of the time it takes a top political team even to agree the points on its reform plan. The impression takes hold that commercial activity is inherently more efficient than the government's. It's a view that the government does its best to encourage, by choosing the biggest state-sector shambles as the first to be spun out to private management. Performance gains are further guaranteed by building into project evaluations an assumption that private companies will achieve lower cost, despite paying more than the state for their capital; and by focusing on the immediate exchequer savings when private funding goes into big projects, ignoring the build-up of longer-term taxpayer costs through private management fees and loan guarantees.

For politicians, spending more every year on healthcare, welfare, education and social housing is a measure of how well they're serving the public interest. They're eternally frustrated that public applause dies down when the corresponding bills arrive. For business, honour lies in spending less on these causes each year: by the terms of its contract, this can only mean that a better-off public is in better health, doesn't need extra lessons and can buy its own homes. The bonus rises further when a smaller tax bill falls on the doormat of the rent-to-mortgage home.

## The Counter-Movement

Anti-globalizers' self-styling as a 'movement of movements' is wisely chosen. Their frequent complaint is of a world that promotes free flows of products and profits while clamping down on free flows of people. Their often favoured solution is letting citizens move at least as quickly and freely as cargoes. Source and destination country are both held to benefit when people are pulled across borders by economic opportunity, before pressures requiring refuge or asylum push them across.

The movement of movements collides with a worldwide political establishment which could well be characterized as a movement against movement. The protesters' preferred globalization, which promotes the international movement of people, works against the past century's pursuit of economic powers and citizenship rights, which largely depend on people staying put. The system of democratically regulated capitalism developed in Europe, North America and East Asia over the last hundred years depends on most people, and their incomes and ideas, staying rooted within national borders. Free trade and free capital movement promote this settlement if they make free movement of humans unnecessary. Commodity flows undermine political stability if – as protesters claim – they set up conflicts and inequalities that also drive people into motion, while demanding that poverty and persecution reach boiling point before the border crossings open.

Political fortunes are made when large numbers shift their ideological ground. Tony Blair's success is that of a youthful radical who took his baby-boom generation with him when he made the usual mid-life slide to the right. At the moment when professional success and financial reward were dulling the personal appeal of state-led tax-and-spend, a whole cohort of post-war socialists were undergoing the same maturing moderation. They seized on a leader who could claim complete sincerity to Old Labour principles, while explaining that tax cuts, private health and selective education, means-tested benefits and workfare were fully consistent with the socializing mission. Across the Atlantic, George Bush Jr rode an equivalent swing on the political right, being there when Republican moralists were finding that second marriages, disease-fighting experiments, genetic modification, daytime TV and other actions frowned on in political adolescence were well worth warming to as political adulthood ran its course.

Political powers are, however, undermined when large numbers shift their geographical ground. Movement in and out of local communities erodes 'social capital', the knowledge and trust contained in relationships among people who talk on first-name terms. Social capital underpins private traders' confidence that goods will be delivered, debts settled and sale-or-return agreements honoured, without exhaustive policing and constant resort to legal sanction, whose additions to 'transaction cost' would rule out a large volume of otherwise profitable trade. Movement in and out of organizations does similar damage to their stocks of 'human' and 'intellectual' capital. Vital skills, task-specific knowledge and team-specific understanding can be lost if people are moved around as part of a re-engineering or laid off in pursuit of downsizing. Workers are less inclined to take additional training, and employers to provide it, when they don't expect to stay on long enough to generate a payback from the investment. Erosion of social and organizational capital risks undermining the foundation of free enterprise and market-based transaction.

Movement in and out of nation states has done much to make good those nations' loss of social and human capital. Networks formed by immigrant groups often prove much stronger and more commercially and culturally productive than those of indigenous people, as exemplified by South Asian small business communities in Britain and jazz pioneers in the US. Injections of skilled migrant labour have been vital in filling spare science degree places and information technology vacancies in richer industrial nations whose own youth talent supply is drying up due to falling birth or literacy and numeracy rates. But even as it fills holes in the system's production base, migration knocks holes in its political base, by eroding the political capital of those who try to uphold redistributive welfare and rights-rich, civic forms of citizenship.

An economic system in which inequality is generated through free trade, and perpetuated through unrestricted property ownership, has two ways to stability. One is to give the losers plenty of chances to turn into winners, through new jobs, training or enterprise opportunities that generate upward social mobility. The other is to ensure that the winners give back to the losers, by choice or by force, enough money and money-making opportunity for everyone to feel better off than under a more trade-regulated, ownership-restricted system. Systems of social mobility and economic redistribution must be big enough to bring all of the winners and losers together, but still small enough to tell them apart. Local jurisdictions, too close to the ground, risk clustering especially advantaged or disadvantaged sub-groups. Global jurisdictions, too far above it, allow the winners to ignore (or pretend to be) the losers. So today's humanized forms of capitalism are almost always built around a nationally administered welfare state.

Legitimation through redistribution depends on a shared sense of social solidarity. Where redistribution is funded through taxation, those who know (because of their inherited wealth or business success) that they will be net payers into the system must feel a sense of identity with those who (through inability to sustain a living wage) are destined to be net recipients from it. In the ideal world of political theory, this identity is built on universal human values – rights to life, freedom from certain constraints, and freedom to make certain achievements within that life – which everyone in principle can share. In the real world of political practice, shared identity almost always rests on relative cultural values which can be shared only by communities closely defined in space and time. Shared nationhood, history, territory, religious faith promote a sense of oneness with fellow members by creating – and often appealing to – a sense of difference from non-members. So, for example, top-rate British taxpayers tolerate a significant percentage of their income being channelled to help less fortunate British citizens, but are happy to see only a tiny fraction diverted to the millions who live in much worse poverty outside Britain.

Legitimation through social mobility depends on 'open' educational, political and corporate structures, through which anyone can rise if they make the necessary efforts. Such mobility has rarely been attained. Studies across the 'industrial' world show that, while children can often make the leap from their parents' poverty into relative privilege, rags-to-riches within a lifetime is much less common. Studies of lower-income societies show similar lack of intra-generational mobility, despite faster rates of growth and organizational change which are usually assumed to make old hierarchies more 'open'.

Where high rates of social mobility are attained, they again seem largely dependent on social solidarity achieved through exclusionary membership. This is because, firstly, personal effort alone can ensure a rise through the hierarchy only if other inhibiting factors (lack of initial wealth, education and health, or just bad luck) are fully compensated for. Compensation requires welfare state arrangements which are usually exclusionary, as argued above. Secondly, upward mobility is only limitedly linked to formal effort and attainment; personal judgements are also involved, and these frequently depend on the sense of shared identity between those who select and those who apply. Companies look for employees who will fit into their 'culture'; professions for practitioners who obey unwritten rules as well as following the formal code; parties for parliamentary candidates whom local voters will regard as 'representative'.

Openness of élites for people who achieve 'integration' into prevailing social norms is attained at the cost of closure against those who are judged to have rejected or failed at integration. These excluded minorities, from the margins of the indigenous society as well as from parts of the immigrant community, get stuck on the lowest rungs of established career ladders, experience the welfare system as a poor substitute for sustained employment rather than as an effective preparation for it, and are at risk of passing down their excluded status through childhood deprivation.

Redistributive welfare states depend on a stable economy, generating enough growth to lift the lower end of the income scale without penal taxation for the higher end, and enough jobs to keep most people free of welfare dependency for most of their working lives. Social mobility similarly depends on a growing, job-creating economy to create continued openings in hierarchies whose early-rising members have become entrenched, and so ensuring that those bettering their skills and experience will have higher places to move to, making their human-capital investments worthwhile. The economic management required to combat cyclical fluctuation and sustain high employment without excessive inflation has traditionally been pursued at national level. This creates a further dimension in which enhanced citizenship rights at the 'core' of a nationally defined society depend on abbreviated rights for those at the 'periphery'.

Although management of aggregate demand, through fiscal and monetary policy, has been central to industrial nations' efforts at economic stability, most have also depended on selective immigration and 'guest labour' policies to iron out successive phases of labour shortage and surplus. Migrants – especially those who brought private wealth and scarce work skills – were invited in to Western European social democracies when rapid growth caused labour shortages in the 1960s, but many were encouraged to leave again – or thrown out of work with minimal social support – when slower growth sent unemployment sharply upwards in the 1970s and 80s. The door to economic immigration is being opened again as a demographic decline, shrinking the workforce through fewer young entrants, leaves Western Europe and North America needing a new injection of higher-productivity labour to support the growing numbers, retiring earlier and living longer, whose pensions – whether financed through public payouts or private investment funds – entail an ongoing transfer of income to those in retirement from those still in work. But while higher immigration enhances the accumulation of the physical, financial and human capital needed for private-enterprise-driven growth, it is widely viewed as endangering the social, intellectual and public capital that provide equally important institutional support for that growth.

## Devaluing Citizenship

The citizenship offered by post-war Social and Christian Democracy in Europe, and Democratic/Republican politics in the US, carried particular privilege through being confined to particular people. Because 'civic' rights (to work, to strike, to receive a living wage, to receive basic healthcare, housing and education regardless of circumstance) carry high treasury costs, they were conceived as applying only to people who did all they could to contribute to those costs. Everyone who could work was expected to do so, in order to create the resources to be redistributed and pay the taxes with which to finance redistribution. Hence the centrality of full employment to the Beveridge welfare agenda in post-war Britain. Everyone offered expanded civic rights was expected to take them up in the right measure. Hence the severe strictures both on those who over-used them (such as unemployment benefit claimants who also worked) and on those who under-used them (such as parents who refused to send their school-aged children to school).

There has been little success in creating a foundation for redistributive norms that does not depend on exclusionary solidarity. One notable effort has been to create a stronger set of universal principles, so that membership of the redistributive group can be opened to all without being diluted into platitude. The inspiration has been that of professional ethics, the set of baseline knowledge and behavioural requirements by which lawyers, accountants and other professional groups define and defend their membership. But two aspects of professionalization fatally undermine the use of this model. In practice, ethical codes have frequently failed to rid the professional group of exclusionary practices. Aspiring judges, doctors and professors from the non-white, non-male, non-'mainstream' communities have frequently complained that they do not receive equality of treatment with 'mainstream' colleagues (from those colleagues or from clients), despite formally attaining the same professional standards. More fundamentally, in principle, even a professional group that does not discriminate against suitably qualified members achieves its solidarity only by strict exclusion of those not suitably qualified. Professional guilds and associations justify their exclusiveness on the basis of upholding standards so that clients are assured of good service. Critics charge that their exclusiveness is designed to restrict entry and competition, so that clients are assured of unnecessarily expensive service.

A second approach has centred on free-market transaction, as a universal forum in which anyone can buy and sell to better their own condition. The market is notionally blind to people's history and heredity, so long as they obey the standard rules for entering and executing trade. But these rules include honouring promises to pay, and abandoning transaction when the bidding gets too high. Ability to obey the market rules depends on ability to pay the market price. So 'equality before the market' tends to magnify, rather than remedy, the inequality in other social interactions arising from unequal wealth and income distribution. Because it rewards successful traders with extra purchasing power, the market tends to amplify initial inequalities. And as it falters under its own exclusionary weight, battle over the division of the economy's limited riches intensifies and the exclusionary nature of economic citizenship gets even more starkly exposed.

In the absence of solidarity through universal behaviour or belief, state-enforced redistribution loses social legitimacy. People no longer expect to be net payers into the system over their lifetime, but instead expect their tax contributions to bring proportionate rewards, and their social insurance payments to hold the same potential rewards as private insurance policies (with which many start to supplement their national insurance). Welfare states' boldest elements have always relied on proportionality of this sort. The middle-class households who pay most into Britain's National Health Service also get the most out, in terms of NHS treatment and the health improvements that it produces. The general drift from tax-based welfare back to social insurance, with clear links from premiums paid to privileges received, shows proportionality becoming more explicit as solidarity declines. As participants once more expect to get all they pay for, the welfare state ceases to be a mechanism for redistributing income between different individuals at a point in time, and becomes a mechanism for redistributing income to the same individual at different points in time.

A regressive solution to the problem of international fluidity undermining national solidarity is to remove the ability to move: by tightening border controls, toughening asylum policy and restricting 'green cards' to those who bring in human or financial capital in return for their right to remain. The more progressive option is to remove the incentive to move, by giving people better chances to acquire and apply commercial and creative abilities at home. Limited success with the second approach has driven social-democratic governments to increased use of the first. New Right alternative governments have a radically different solution: to exchange strong civil rights for much weaker basic human rights, which all citizens can share and all states can afford. States which stop at guaranteeing the right to live, speak freely and seek work can be much more generous with their membership than states which go on to offer rights to live at a decent standard, be listened to, and actually find a steady job. So governments that fight inequality and exclusion within their nations are seen to promote and perpetuate them between nations. The radical right, with its slimmed-down citizenship, captures the internationalist agenda, and it is for left-leaning governments that tilt in this direction that anti-globalizers reserve their strongest fire.

## The Privatizing Mission

Many of the powers once trusted to governments – to resolve disputes, solve collective action problems, regulate the economy and its natural monopolies, defend the realm and stop the present generation from mortgaging its children's future – were assigned on the assumption that rulers served a distinct and defensible 'public good'. Politicians and public servants were supposed to subordinate their own interests to those of their electors, promoting other people's interests as the only means of preserving their own.

In most market-driven countries, such superhuman neutrality is no longer considered either possible or desirable. The market mentality increasingly takes pay as an index of what people are worth. So staying in a low-paid public-service job seems in itself to affirm the worthlessness of bureaucracy – though when top public servants are found to have leapt into the tycoons' pay league, a seamless shift to 'fat cat' accusations keeps the condemnation intact. Big government's opponents assail it with a double-sided argument. States can no longer afford to do the things that might be useful. Nations can no longer afford to let them do the things that aren't. If a job is worth doing, the marketeers insist, private enterprise can do it. Or if it can't, the damage wrought by tax-and-spend means that it's better left undone.

Pessimists from the self-styled 'public choice' school insist that this seepage of central power is permanent, not the latest replay of a regular cycle of ebb and flow. The cyclical view sees government, like business, as continually devolving and then reclaiming power, in response to change in the demands placed on them and the people placed in them. Responsibilities are spun out when the national government needs to relieve a bottleneck in (and share the public dislike of) decisions piling up at the centre. They are taken in again when squabbling shires and recalcitrant regions prove incapable of reaching any decisions, except to go back to the centre for more funds. This alternation of policies tends to move into phase with one of personalities. New bosses, determined to stamp their mark on the organization, set out to change whatever system is there, so that those who inherit a centralized system will announce grand plans to diffuse authority downwards, and their successors instinctively reach out to pull it back up.

Companies go through much the same cycle: devolving power when operations need a ground-up redesign, taking it back in when quick thought and urgent action are essential at the centre. As in most areas, there is a difference in the wavelength of the cycle. Commercial vehicles can do the U-turn in a matter of weeks, while political power-steering can take years to turn full circle. People get edgy about democracy, or impatient for their poll-night excitement, if electoral cycles last longer than four or five years. But so short a term scarcely allows politicians to draw breath, or change anything meaningful, before victorious first-day promises of a heroic U-turn give way to mid-term laments at how long it takes to turn a supertanker around.

## No Longer in Demand

Public provision now resembles a last resort when it comes to managing the supply of vital services. But even more spectacular has been the dumping of a government role in managing demand. When the world emerged from its 1930s slump, armed with a new set of theories of how the macro-economy worked, it seemed that public spending had found a new role at the centre of the private economy. Markets sometimes failed to work in synch, yielding insufficient spending power to match industry's producing power. The public purse then had to open, to cover for private wallets that poverty had left empty or precaution welded shut. To politicians' elation, it wasn't always necessary to raise revenue before boosting state expenditure. At times of downturn, there was positive advantage in spending what the Treasury hadn't got. The deficit would be made good later on, and public borrowing paid back, from extra revenue once the kick-started production base got back to full capacity.

Even better, given the crossfire of lobbyists that breaks out whenever politicians get out the credit card, it didn't really matter where the money went, as long as it was spent. John Maynard Keynes, who identified the market economy's demand-deficiency, and had noticed a matching imagination deficiency in the public bureaucracy during his time in the civil service, jokingly suggested that digging holes and filling them in again would do the kick-starting trick. What mattered, given the mass unemployment with which the 1930s began, was to inject some extra spending power so that the idle hands could go back to work on the idle machines. Equally desirable, given the mass rearmament with which the 1930s ended, was that governments at least spend on something that didn't destroy, if they couldn't alight on something that might actually create.

Contrary to what half a century of balanced budget believers claim, Keynes isn't dead. He's just been rather restricted to the countries that still offer him asylum. The US, with its self-confessed duty to stay rich so that the rest of the world can go on feasting on its breadcrumbs, has no hesitation in reaching for Keynesian solutions when its own economy stumbles. The Bush tax package of 2001 'gave away' a fiscal surplus which had already disappeared due to missed growth targets, swinging the federal budget into a deficit not matched for two decades. Since, in fact, the early 1980s under President Reagan – another Republican who preached fiscal rectitude but dived for the deficit when recessionary rocks approached.

But Keynes' writ is not allowed to run to countries that take their economic cue from the US, through its influence on the IMF and other policy-approving institutions. There, even the smallest deficit at the lowest moment of the cycle is a cue to cut state spending and raise taxation, despite the invariable result that the demand gap widens further and recession is prolonged. Balanced budgeting, low inflation and a stable exchange rate are deemed essential to keeping lower-income nations open for banks and business, even though these measures tend to depress activity further, and undermine public support for long-term measures to cure the crisis tendency.

Economists advising the multilateral lenders-of-last-resort have tended to concede this after seeing the human impact of the standard 'adjustment' formula, most recently after the near-collapse of Indonesia in 1997–8 and Argentina in 2002. But faced with a choice between the cost of economic prescriptions that might need much greater transfers of aid and cheap credit to poorer nations, or of running another job advertisement, these lenders' shareholders tend to prefer shooting the messenger to reading the message. The last World Bank chief economist to express well-founded doubts about deflationary adjustments, Joseph Stiglitz, was sacked from the post for misreading its politics even as the worth of his economics was about to be rewarded with a Nobel Prize. The European Union, showing some awareness of austerity's limitations, allows its members a degree of fiscal 'activism' to defend their growth when external trade and monetary conditions are unusually tight. But its monetary union constrains members to keep state debts and deficits down to well below the scale needed to contend with a serious downturn. Even a 'golden rule', limiting the fiscal deficit to what the government is investing, upsets those who feel that such capital projects should be left to the private sector.

Keynesian solutions work, it is explained, where governments are capable and honest enough to make the right deficit spend on the right things at the right time. But the quality of government is vital for such intervention to work. A politician like President Bush, elected with a large majority on a perfectly transparent vote with no hint of financial favours from powerful companies, is obviously completely competent to spend his nation into deficit. Elected authorities in places like Argentina and Indonesia are evidently made of less trustworthy stuff.

Retailers of such explanations are right to suggest that what is appropriate for America may not work so well in other places. They are almost certainly wrong to write them off there, or suggest that the difference in prospects is due to contrasting political strengths. America can spend its way out of recession because other nations will lend it the money. If they don't, its citizens can lend it to themselves, making use of America's reserve currency privilege and its ability to secure borrowing against financial assets whose value is further boosted by the credit growth. Japan can also (and regularly does) spend America out of recession, running fiscal deficits which – because its public saves as assiduously as its public servants spend – flows straight out into demand elsewhere in the world. Smaller, poorer nations can generally no longer do so, because deficits set off economic signals that counteract their expansionary force. Interest rates rise, restraining the private sector's investment spend as fast as the state expands its own. When investors also fear that bigger government is the prelude to more inflation and increasing regulation, a fiscal 'stimulus' may even cause activity to fall.

## Giving Money Away

Public debt was one of the earliest forms of money, and government was the borrower on which many of the world's great banks grew rich. But along with abandoning fiscal control in favour of the balanced budget rule, most governments have surrendered monetary control to an independent central bank. Having dismissed politicians' competence to decide what to do with money, it is unsurprising that government's detractors should want to prise their grip away from the source of that money. The thinking is that it takes strong and impartial minds to determine what quantity and cost of money will ensure price stability and output growth – and to stop the state's dead weight from undermining that calculation.

Left in control of the cash till, governments can – and usually have – printed the money that they don't dare demand from taxpayers, setting off an inflationary spiral which upsets the economy's income distribution and chokes off its growth. Even if governments were able to avoid this temptation, the constant fear of their yielding to it saps private sector confidence, and ties the market's 'invisible hand' to an all-too-visible handbrake. Only a central bank which announces its interest rate setting rules, or acts consistently so that private investors can work them out, is capable of creating the stable decision-making conditions in which consumers get the income that underlies their spending plans, and industry makes investments that deliver the expected rate of return.

## Democracy on Demand

A perennial lament of the political left is that if voting changed anything, they'd abolish it. 'They' were once the political right. Increasingly, the ballot-snatching impulse strikes hardest at reinvented Labour and Social Democratic parties. Free-marketeers used to fret that free votes would lead the public, focused on their interests as consumers, to foist unbearable regulation on already overburdened producers. They also viewed the political arena as natural ground for an extension of the division of labour. Politicians, where needed at all, should rise to their posts on the basis of special competence in dispute-resolving decision-making, not through the more conspicuous but less constructive ability to mislead the electorate.

True to old socialist suspicions, pro-marketeers' discovery of the uselessness of government has given them a new enthusiasm for democracy. Pro-globalizers are proud to point out that the lifting of commercial and cultural barriers has been accompanied by the spread of elective democracies. Of around 190 sovereign nations at the turn of the century, well over half could discipline their rulers with a degree of voting power. Simultaneous expansion in the representativeness of governments, the level of integration between their economies and the lack of engagement between their militaries fuels the 'neo-liberal' argument that *laissez-faire* economics is the bringer of democratic politics and international peace.

The link between free markets and free elections is argued to run via decentralization – of control over resources, and of the power over people that this gives rise to. Economic choices build *up* from the independent actions of scattered traders. Political choices are passed *down* by central leaders, after hearing just a small group of voices, in their cabinet room or in their heads. So the dispersal of resources forces leaders to listen to their people. To have a tax base from which to fund their activities, governments must give businesses what they need. To get the taxes paid, governments must do what people want. The control of resources directly by the state – even if a well-meaning response to 'market failure' – becomes a recipe for overriding the popular will. But people can prevent this, by not paying taxes at all if denied an opportunity to vote them down. In an internationally integrated world, business can move wherever taxes, regulation and infrastructure are most favourable. So states are forced, by competition in the political market-place, to minimize the diversion of resources to the public sector and maximize the efficiency with which they are used.

Such argument neatly chops off an awkward twist in the previous free economy/liberal policy tale. This portrayed democratic government as an outgrowth of the free market, but one which turned around to strangle its creator. The problem was that, because of the need for efficiency, it wasn't possible to share out economic resources to everyone. Some had to have more than others: as an incentive for making risky investments and innovations, or as reward for their special ability to make the wheels of commerce turn. As differential success concentrated resources into fewer hands, the 'have nots' would start to outnumber the 'haves'. Democratic voting would then lead to majorities seeking to redress economic inequality through the political process. Minimum wages, redistributive taxation, collective ownership of major enterprise, unemployment benefits and other interventions would be made to 'humanize' the market. Neo-liberals regard such 'social protections' as counter-productive, though well-intentioned. They undermine efficiency incentives, and choke off the growth that can ultimately raise living standards all round. In the end, equal shares of a shrinking cake are worse than crumbs from one that continues to expand.

Democracy has not, in practice, turned around to bite the invisible hand that feeds it. Electorates have, if anything, shown an increasing propensity to vote for subdued and self-deprecating governments. From America's Reagan and Britain's Thatcher to Brazil's Cardoso and Japan's Koizumi, political strength has become identified with telling people what government can't do. Success is measured by how far taxes can be cut, how much regulation can be swept away, how many enterprises sold off and agencies spun out to private management. Once people grow accustomed to government only stepping in when all other solutions have failed, progress comes to be measured by less, not more, state intervention and expenditure.

## Buying Votes

'Neo-liberal' enthusiasts for non-intervention view this as a vindication of their free-market creed. People vote to protect commercial freedom from political interference because everyone has a stake in the private enterprise economy. Their wages depend on the survival of an employer, whose plans they must therefore support. This is reciprocated, since companies must train staff to extract continued product and productivity improvement, so giving them skills that make it more important to retain them. Their retirement savings rest on the performance of shares, whose steady accumulation by pension funds means that workers are slowly buying up the firms they work for.

Politicians of an earlier era resisted the suggestion that governing best means governing least. The lure of long ministerial holidays was one of the factors that won them round. Equally powerful was the discovery that doing less could be more lucrative, in wage-bill as well as work-load terms: today's non-meddling politicians are rarely paid less than their interventionist forebears, some pocketing more for the burden of doing less. But it would be unfair to accuse our leaders of switching to a *laissez-faire* ideology to make life easier. In practice most must work just as hard as before – to reverse all their predecessors' reforms.

Until it actually devises a better way, politics can only create the illusion of progress by constantly reviving past approaches, and replaying the cycle of their rise and fall. One governmental generation, in the 1950s and 60s, built up the role of the state in the economy. Its successors in the 1980s and 90s knocked down that role, every bit as energetically. Intervention is likely to be back on the agenda within ten years, as a new breed of rulers seeks a new, more active role. The 'minimum state' is not a clever trick by leisure-seeking leaders – it's just the latest, crash-dieted embodiment of a see-sawing approach to a workable interventionism/abstentionism balance.

*Political* democracy is weighted towards the supply side. When it comes to deciding policy, people look to their leaders as producers. Few have a detailed wish-list outside their immediate area of interest. They expect their would-be leaders to parade the manifesto and submit alternative options for the voters' 'yes' or 'no'. When it comes to implementing and financing the chosen policy, people cast themselves in the producer role. Most are more sensitive towards what's subtracted from their individual income and expenditure than what's consequently added to collective provision. The 'social charge' of higher tax weighs more heavily than the 'social wage' from higher public spending.

So when the state has resources to spare, most voters prefer reclaiming them through tax cuts than leaving them for the government to assign. This is partly a matter of control. The incentive to spend wisely is greater when the money is our own, not someone else's. But there is an important range of outlays which, through better coordination and spreading of costs and risks, are much more efficiently made by the state on behalf of individuals within it. Electorates' keenness to cling to 'their' money, even when individual efforts at transport, education, healthcare and pension provision are palpably less coordinated and cost-effective than the publicly-pooled equivalent, attests to the way in which producer-side individualism prevails over the range of choices which governments still take as their own.

*Economic* democracy is, in contrast, weighted towards the demand side. When people vote with the purse, they expect to play an active role in shaping the product they're spending on. Political citizenship means vesting our 'producer sovereignty' in the state, which we expect to appraise and uphold our interests. But we expect to exercise our 'consumer sovereignty' ourselves, rejecting as wanton interference any attempt by political authority (or cultural critics) to tell us what to think or do.

Economic choices can be as hurried and hard as political choices, and there are times when people are as keen to opt out of consumer choice as of citizenship choice. Some decisions, like whether to buy jam today or try the extra-whiteness soap powder, are so trivial that deferring to a specialist would save a lot of time. Others, like whether it's worth trading the car in or buying digital TV this year, are sufficiently serious that expertise could save a lot of money. We don't want government to determine these things for us, in the way that it decides how to reform the primary care system or launch a limited war. But we don't want to incur the time penalties of properly handling the decision ourselves, or the cash and blame penalties of rushing through it to save time. The large corporation steps into this ambiguous terrain, as an agency that can help us to make the choice at the same time as seeming to multiply that choice. Dell helps us to build our own computers without knowing what's in the chips or how they're wired together, and without losing the scale economies that bring it down inside our price range. Kellogg's gives us freedom over the form and flavour of breakfast cereal, while steering it into the shapes and tastes that their production lines are tuned to. In buying (as more than half of us now do) our main consumer products from big store chains, our selection for ourselves is neatly embedded within, and given front-of-shop prominence over, their behind-the-scenes selection on our behalf.

## Getting What We Pay For

The individual benefits from corporate action tend to strike us much more sharply than its largely shared-around costs. So we welcome advertising as a subsidy to newspaper and TV subscription prices, ignoring the margin that gets added to the product price to pay for the ads. We are liable to see a free gift with the purchase as more valuable than the price cut that could have been offered in its place. Consumers' keenness to spend their own money used to lead to inadequate provision in areas where pooled buying power was a route to more or better products. Even when governments took on the responsibility for schools, roads, hospitals and other 'public goods', they could never raise enough taxes to fund them on the scale the public wanted, because self-focused economic votes spoke louder than socially-focused political votes.

Now, with big business bringing organized capability to individual action, people can acquire their public goods through comfortably private channels. The privately managed schools, for-profit hospitals and personal pensions now spreading from America to Europe do more than offer brighter logos and chattier helplines than their public-sector predecessors. They re-establish a link between the service people receive and the subscription they pay, in a way that older forms of state provision or social insurance deliberately did not do. Tax-funded healthcare, education and pensions recognize that, to give everyone the help they deserve, the best-off must pay in more than they ever get back, so that the worst-off can consume in excess of their contribution. Subscription-funded hospitals, schools and retirement plans – and, in the same vein, fee-based television, door security or court services – restore the link between in and out payment. With entitlements dependent on contribution history, those with extra money regain the chance to buy extra privilege. The total cost of provision rises, with scale and co-ordination eroded as private providers spring up beside the state's. But the individual cost falls for the many who no longer have to pay a premium to cover extension of the service to those who draw from it more than they can deposit with it.

Government's critics are right that, in current conditions, most nation states lack the power to control their economic fate. What they conveniently ignore is that governments have chosen those current conditions. If public hands are cuffed, it is because a state-appointed jailer threw away the key.

Discretion over fiscal and monetary policy largely disappears when governments take the brakes off capital mobility. This was reviewed in chapter three, and has largely been the direction of international policy in the past 20 years. A nation that lets capital flow freely in and out must set monetary and fiscal policies to stop excessive inflow (which can strengthen its exchange rate and price its producers out of foreign markets) or excessive outflow (which can drain it of investment funds, and weaken its currency until production inputs become unaffordable). This means setting interest rates at whichever level will avoid sharp downward movements in the currency, or steady upward movement in prices. It means setting fiscal policy to balance the budget over the cycle – a test that no government has yet passed consistently, since doing so would make the cycle disappear.

What keeps an economy steadiest isn't always what makes it grow fastest. Studies of economic expansion have shown a substantial positive contribution from services traditionally provided by the public sector. Tax is not just a drain on private income and expenditure, or a well-meaning redistribution device whose disincentive effects accidentally shrink the cake that it attempts to share out. It also pays for certain items which private enterprise requires and cannot self-provide. Many independent economic studies now show that healthcare, education and basic national insurance systems not only have high private returns – benefiting individuals through higher income and longevity – but also higher social returns, boosting the rate of economic growth. Technological progress, the other great expansion joint in the bridge to higher output, also owes much to public outlay on training and research. If they insist that tax must be kept low, and the public budget balanced, free-marketeers must show that enough of these 'public' goods can be provided by private enterprise. If not, the government's need not to spend more than it takes in tax will hold down growth, as vital social and public capitals go missing from the scene.

But the latest studies also show a further key contributor to long-term growth. This is stability: the social stability delivered by the state as central conflict resolver, as well as the economic stability achieved by government as spender and saver of last resort. As the previous chapter sought to show, market economies left to themselves are very far from stable. Government has a role in supporting private sector growth, by filling gaps in what private producers can supply and what private buyers can demand. For this it needs sufficient resources also to support itself. There is an expanding cost to excessively contracting the state.

Success for the new, business-friendly politician seems to lie in gradual self-abolition. Governments have accepted commercial rules, and joined the market players when unable to beat them at their own game. Most public industries and agencies are being sub-contracted to private management, if not fully privatized. 'Quasi-markets' or 'benchmark' disciplines are being imposed on those that governments cannot sell. They subject remaining state assets to process re-engineering, total quality management, culture change, and a bevy of other buzzwords credited with galvanizing the private business world. They promote 'value-for-money' in place of traditional measures of public service quality, gauging an operation's success by the sums it would raise if private shareholders were ever asked to bid for it.

Private managers are parachuted in to take over the reins from public servants, their salaries multiplied to match.

After years of complaint that bureaucrats brought disaster by telling businesspeople what to do, miracle cures are now claimed from reversing the command chain. Running an expanding soft drinks business is the obvious grounding for revitalizing the public hospitals. Where a good track record in trouser retailing arrives on the city hall's doorstep, renaissance of state school services cannot be far behind. But private profit motives have not been easy to align with public service motives. Economic efficiency, as well as most conceptions of social justice, continue to react against a system that sells educational opportunity and healthcare to the highest bidder, and lets lifetime income alone determine when and how well older employees can retire.

Before private industry is invited to step across to run schools, hospitals and prisons, it might at least be wise to check its performance on home soil. But the leap into privatizing public and welfare services is being taken just as earlier public industry privatizations are beginning to unravel. When energy, telecommunications, transport, banks and other 'strategic' industries were privatized, pro-marketeers accepted that there would be a need for regulation until once-monopolized sectors could be fully opened to competition. The recent collapses of Railtrack in the UK and Enron in the US have shown the limitations of regulation, with private shareholders withdrawing their support unless state hands are removed completely from the reins. The drift back towards private-sector monopoly in telecoms, software and other 'network' industries when those reins are released has shown the difficulty of replacing regulation with full-fledged competition. 'Natural' monopolies turn out to be more natural, and more resilient to a globalized market, than the privatizers gave them credit for. Renationalization is not an option – but only because the writ of the new private monopolies now runs far beyond even the lengthiest national borders.

Efforts to revive state enterprise under private management have rarely gone according to plan. Of the showcase privatizations with which Britain wooed the world in the 1980s and 90s, British Telecom is now dismembering itself to pay off crushing debts, while a nation promised broadband abundance still struggles with erratic dial-up connections. British Airways' profits are in reverse thrust, after it upgraded to business class just as even the tycoons were switching on to budget travel. Even the world-renowned BBC must keep its investigative news staff away from the bizarre paper trail of its internal market, which can make it cheaper to bus an interviewee to central London than let them walk to their regional studio over the road. British Rail's successor is bankrupt, while the private train operators working through it often have more staff deflecting their compensation demands to three different regulators than dispensing new timetable information to stranded commuters awaiting the 5.15 from Crewe.

At least, however, the lights stayed on when Britain privatized its power supply, even though this was largely because of careful peak-time planning by the previous Central Electricity Generating Board. When California tried a similar sale, the public's dislike of cooling towers in its backyard – and the taxes to pay for them – overpowered its preference for more power. Unwilling to pay the market price for scarce supply, they ended up having it rationed. The parallel with Soviet times was intriguing, but misleading. Russians were by then enjoying 24-hour power, their problems in restructuring the main generator having prevented the economists from wresting the on-switch from the engineers.

Regulators are a large part of the public–private problem. Shareholders take fright at holding stock in companies whose prices and investment plans must still be approved by arms of government. Yet utilities which diversify abroad, or into new business lines, to escape the regulatory shadow, clash with investors' preference for keeping the risk-spreading to themselves. Private water company management could not make their outfit a going concern, even in a country as precipitation-prone as Wales. Glas Cymru, at least, pro-actively made itself a non-profit-making company. For state-sector spin-offs such as phone firm Energis and nuclear waste reprocessor Sellafield, renunciation of returns wasn't even a matter of choice.

But unregulated privatized companies, free to price and spend as they like, still can't always make the numbers add up, even with the state continuing to slip them top-up subsidies. Britain's privatizing 21st century opened with successors to British Coal and British Steel performing a downsizing duet on the Industrial Revolution's birthplace. If public ownership ever needed a beyond-the-grave tribute, it can probably find one in the Range Rover: archetype for the up-market all-terrain vehicle, and a creation of the British Leyland Motor Company, 30 years ahead of its time.

Private management initiatives within the public sector have been equally disaster-prone. The internal market in Britain's National Health Service produces queues for operations longer than those for coffee rations in the last days of Gorbachev's Russia. Private financing of state schools has meant confectioner-sponsored health information and a Coke machine in every canteen, but not the flood of unconditional funding which was supposed to replace the pre-metric textbooks and repair the leaking roof. The security firm that manages a refugee reception centre sues the local police when inmates riot and burn the place down. The only consolation is that those who worked on these disasters have plenty of time to right the damage – because their market-linked private pensions can't possibly deliver enough to retire on.

While politicians feign shock at needing marriage guidance for their public–private partnerships, it should be no surprise that firms can't recreate triumphs in their own narrow market at the general government level. Companies can set their mission and then choose their products, customers, workforce and suppliers to suit. Governments must get their agenda approved by, and its implementation and financing accepted by, whatever raw material elects them. And while companies in crisis can cut their losses by closing plants and reducing payrolls, that swelling of the private-sector scrap-heap is where the woes of cash-strapped government take off.

## None of their Business

Fans – and foes – of MNCs like to point out that Wal-Mart, with less than a million employees, sells as much in a year as Pakistan, with a population running towards a billion. By implication, submitting to direction from the world's biggest stores group would be a small price to pay for getting national productivity quadrupled. But the comparison misleadingly mixes two size measures. A company's annual revenue would only compare with a country's GDP once all expenditures on materials, marketing and other sale costs were first subtracted. When GDP is compared with corporate profits, countries push companies far down the giantism league.

More importantly, the parallel ignores the way in which multinational companies can filter out their members, a privilege that national governments have hardly ever enjoyed. Hitler sought to annex those he thought would raise the Third Reich's productivity performance, and eliminate those he thought would not. Stalin redrew boundaries and deported minorities with the aim of sifting communism's heroes of labour into higher concentration. Their fate has, fortunately, deterred most later dictators from similar electoral self-selection schemes. While employers can cream off the best and abandon the rest, a state must look after all of its people – even those whom it imprisons, and those who scorn its offers of education and employment. When companies dismiss staff that they can't or don't want to keep on, the state takes them on, if no other employer will. A firm that took on large numbers of employees, and then threatened to go out of business, could expect plenty of help with its debts and investment needs. Governments whose debts get unmanageable or whose client base evaporates have no such ready rescuer. Business could never provide for, or give jobs to, all of the people that a state must look after. A commercial grounding gives no inclination to, nor any experience at, doing so.

Governments' desire to hand past responsibilities to private enterprise arises more from loss of old political courage than an upsurge of new economic conviction. Politicians have recognized that – against welfare-state architects' original predictions – public appetite for education, law and order, transport, health, pensions and welfare protection, environmental safeguards and affordable housing rises more than proportionally to national income. They are disinclined to break the news that, if expanded amounts of these are still to be provided publicly, more of the public's income must be taken in tax. So instead, they reach out for any think-tank or consultancy report which hints at how such outlays might in future be funded privately.

Recognizing that people must buy 'public goods' privately if they won't give government enough to pay for them, left-leaning parties with traditionally big tax-and-spend commitments have become the world's most enthusiastic privatizers. In the UK, a 'Labour' government entered the new millennium with plans for private companies to build and run schools, do relief work for the National Health Service, reconstruct the rail network, even provide air traffic control and supply mercenaries to supplement an over-stretched army.

Private partners don't always seek to profit from the public services they get involved in. The world's stock of football clubs, national newspapers, film studios and under-sea tunnels would be significantly diminished if the profit motive always overrode the sponsors' personal vanity. But in general, private firms need new incentives to take on these roles, and resist old instructions on how to perform them. So surgeons start charging appearance fees. Snackfood-sponsored schools serve up educational courses between doses of the slightly more edible variety. And those travellers who still have a train to catch must watch it pull away as the ticket clerk stops to sell bread, sweets and newspapers to those ahead of them in the queue.

These steps to devolve existing public responsibilities to lower administrative levels are being made just as new responsibilities descend on national governments from above. Potentially dangerous interactions in the global environment, financial markets, trading system and health situation require urgent collective solutions: to coordinate counteracting policies, and to provide a new range of 'global public goods'; ranging from rainforests to anti-meteor shields and international commercial and war-crime courts, these are services and dispute-solving arrangements which every nation needs and no individual nation is willing or able to provide.

## Rebuilding from the Top

Globalization undoubtedly forces governments to scale down tax demands that might scare private enterprise to other shores, and step up infrastructure provision of the sort that investors might need to persuade them to stay. The resultant contradictory demand, to spend more and tax less, is inevitably one that politicians wish to sidestep. So apart from providing antiglobalizers with their most exciting target practice, today's global governance institutions – the IMF, WTO, World Bank and other vexed acronyms – serve two main political functions. They allow national governments to excuse themselves for imposing unpopular tax increases and spending cuts, by announcing that the multilateral rule-setter has forced them to do so. And they offer a periodic weekend retreat, where leaders can meet to commiserate on how impossible the demands on their office have become.

Successful global integration requires them urgently to add at least four more functions. These are mechanisms for stabilizing international product and financial markets; for redistributing global income; for supplying the missing or under-provided international public goods; and for keeping markets working, by eliminating barriers and protecting property rights.

The choice of how to distribute these powers between national and supra-national agencies is hard, because there is a delicate interaction between them. Governments must maintain stability in their own economies, and harmonious relations with other economies, to create the climate in which multilateral agencies can attain and exercise power. Those agencies must help governments to broaden the range of their macro-economic management, redistribution and public goods provision to take account of new cross-border connections and spill-overs. It is a tall order, and one for which time is uncomfortably short, making the smallest disruption unhelpful.

Their heritage may be unfavourable and their structure deeply flawed; but as these agencies are descended from an earlier, more internationally-motivated time, it might not be possible to invent them if they did not exist. And as all organizations capture intellectual and social capital of some value, it is preferable to reinvent them than to knock them down and start all over again. Despite (perhaps even because of) self-confessed past failures, the World Bank is still the best available route for channelling cheap loans to countries that need more capital, the World Health Organization the best forum for devising and distributing common disease cures, the WTO the best hope of getting rich countries to renounce the protection that they deny to others, and G8 the best mechanism for stopping the major economies' policies and practices from colliding.

The need for reform is recognized even within these organizations, and is already being tackled. The World Bank and IMF are at last working on a bankruptcy plan for nations that would give them the same work-out opportunities as a private creditor. The WTO is working to get its disputes panels free of over-powerful, and still over-protective, rich-country members. These and other initiatives must be broadened and speeded up, not sabotaged. The tragedy of anti-globalization demonstrations outside international summits is that they may be hampering the very processes their activists say they want to help.

## CHAPTER SIX

# Out of the Loop

*Art, culture and communication haven't made much of an appearance so far. The finer things in life are forced to fight for their place in a world linked primarily on economic terms. Globalization risks driving the world towards monotony and monoculture, as economic integration and cultural standardization stamp out alternative lifestyles and learning. Like all monocultures, the winning formula will flourish and then die – if not of boredom, then of inability to adapt when natural or social change call for alternatives that no longer exist. But arguments linking commerce's onward march with the dumbing-down of culture miss a more positive interaction, through its expansion of network technology. This not only quickens escape from the tyranny of localization, but could also be the basis for a growth that trades abundance of material for a broadening of the mind.*

Globalization isn't all about capital, commodities, commerce, and the other things economists like to spread gloom about. Life's more uplifting activities are also going global. Their success in doing so will play a large part in deciding how long the economy stays that way. For capital's more cultural forms, as distinct from its economic starting-blocks, international travel tends to broaden the mind. The problem is to let unbounded culture re-shape the borderless market, not the other way round.

When music, art, literature, scientific knowledge and ideas cross borders, they tend to improve understanding between peoples, while generating novel styles and new discoveries that add to cultural value. This is in contrast with the trade disputes that frequently erupt when goods and services venture abroad, and which generally destroy economic value. But culture's expanding horizons are in danger from these economic conflicts. The tension between commercially reshaped culture and a culturally reorientated commerce is potentially both globalism's gravest, and its most creative.

## Plutocrats and Philistines

One of the biggest complaints about the present direction of world development is that music, art and literature are being turned into commodities, ideas and knowledge commandeered as forms of capital. Their intrinsic interest is drained, and expressive merit compromised, once they are harnessed to the task of making money.

Industrialization's original promise was to speed the fulfilment of material needs, releasing time and energy for producing and consuming the finer, more ethereal things in life. The dullness and repetitiveness of simplified, mechanized jobs was rarely denied. Instead, the pledge was to get them over more quickly. Ever shorter, but more productive, stints of commercial production could then make room for ever longer, more enjoyable spells of cultural consumption. Playhouses, galleries and picture palaces would spring up around the grim factories, eventually taking their place as automation ended the need to dirty hands on bruising production lines before wielding a paintbrush or strumming an instrument with them.

This dream is punctured by a market system whose stability depends on growth, through the mechanisms outlined in chapter four. Caught on the treadmill of financing today's consumption from investment in tomorrow's production, our preferences become shaped around whatever is most reproducible. Artistic creations remain on the sidelines, except where industry can add them to its production lines. The transformation of the mass-produced soup tin into artwork was achieved as early as the 1960s, but at least Andy Warhol took the trouble to make a painting of it. Now the label itself is the exhibit, as art becomes a by-product of attention-seeking packaging. A walk down supermarket aisles replaces visits to the modern art museum – which responds by merging its display space with the gift shop and restaurant, so those who see no merit in the paint can look for poetry in the price tag or sculpture on the plate.

When commercially driven, cultural globalization crushes the artistic muse under autistic market force. On the demand side, workers forced to spend most waking hours on the job must streamline their artistic activity, which gets shunted to the sleepiest hours. Just as satisfaction of nutritional needs is sped up by the drive-in restaurant, and health needs by the round-the-clock pharmacy, artistic and knowledge needs must be met from a truncated leisure budget. This means keeping products simple, succinct and rapidly absorbed. So pictures must be instantly enjoyable. Text must amaze or amuse on first reading. School lessons must match the ready-meal's five-minute preparation time, recapturing imagination from comic or soap opera. A liking for Prokofiev or Proust is no longer a feeling that deepens over time, but a fad that must vie for continuing attention with the competing claims of ephemeral poetry and sell-by-dated pop.

So on the supply side, painters, poets and programme-makers must cater to shifting and impatient tastes. They must innovate to make earlier work seem insincere or incoherent, rather than build it into a consistent and developing picture. They must research the likely viewers and listeners just as thoroughly as a business evaluates its customers, and market themselves and their wares as aggressively as sellers of bags or bricks. Improvements in managerial planning techniques now make it clear how reckless Tolstoy was to invest so much in the single, highly uncertain *War and Peace* project, and how foolish of Beethoven to invest so much in only nine symphonies, when a succession of shorter and easier pieces would have achieved a much more sensible spread of risk.

## Matisses for the Masses

The gallery-goers' charge against globalized artistry mirrors that of eco-warriors against a globalized economy: it has taken a time-honoured stock of art and knowledge and replaced it with a fleeting and forgettable flow. Impermanence must be built into commercialized culture, like obsolescence into commercialized craftwork, so that the old can quickly be cleared away to make space for the new. Where artworks' value persists, because of timeless form or famous name, ways must be found to keep recycling them through the market: putting the Rembrandts up for auction, re-mixing the Mozart minuet, reviving the Shakespeare play.

As well as its green leaning, there is a socialistic streak in the current war of culture against commerce. Market pressures are accused of de-skilling the work of artists in an effort to make them produce more; they must then de-skill the process of artistic consumption, so that eyes and ears can cope with all of the cultural creations now flowing their way. The broadening of artistic exchange from a tiny élite to the total population is not the progressive move that its promoters like to present. This is because, like any commodity, the product devalues on multiplication. Just as cheap, mass-produced cars and crockery are not the Bentleys and Wedgwoods of old, so – connoisseurs are convinced – great literature loses something essential when pared down for peak-time TV dramatization, and classical music in conversion to CD.

The loss of quality is bound up with one of variety, since commerce has problems coming to terms with artists' endless experimentation. Once industrialists have designed a new product and seen that it passes the market test, their aim is to sell as many identical copies as possible. There is a 'fixed cost' in getting to the prototype stage, which needs to be spread across the subsequent production run. Much effort is then made to refine and subtly redesign the product, so that people will decide their old version needs early replacement. But major change is discouraged, because making it incurs another big fixed cost, and launching it would risk causing the existing market to dry up. This industrial ethos, of relentlessly refining an existing product but resisting big changes of form, is precisely the reverse of the artistic ethos. Once artists have seen a style, shape or substance become established, their automatic impulse is to stop it there and tackle something new. Fame lies in the fixed cost – being first to discover, leaving others to develop.

This reversal of priorities is common because industrial products are rarely at their best on their first appearance. Whereas with cultural products, the original is usually the best. Commercial fortune generally goes to the 'second-movers': those who watch others incur all the costs of thinking up and rolling out a new idea, see how they could make it better or cheaper, and launch their improved version into the market that someone else has opened up. Toyota did not invent the car, Dell the computer, or Citigroup the bank. They just became unrivalled refiners of someone else's idea. Artistic fame, in contrast, generally goes to those who get there first. Few have ever got very far with the argument that improvements could be made to the Brontës' novels, Eisenstein's films, or even the Beatles' ballads: because everything else in that line owes a debt to their inspiration, and satellites are not allowed to eclipse the stars from which they came.

## Cargo Culture

If art's globalization through commercialization were genuinely making it democratic, then metaphysical poets would be topping the bestseller lists and orchestras packing out the football stadiums. But creative genius is naturally scarce, therefore not reproducible as the system requires. So it has to be toned down before being opened up to the mass. In making this complaint, connoisseurs put other people's money where their mouth is. But this is inevitable in societies whose arts administration is assigned to masters of business administration. In the trading of culture, as distinct from the culture of trading, those who pay the piper don't expect to call the tune.

In the commercial world, a world of prices, those with the most expensive tastes are also those with the most exorbitant incomes. So the discerning motorist can still rise above the crowd by overtaking the Skoda in a Rolls, and the lover of fine food stride past the Wimpy on the way to the Savoy. But in the cultural world, a world of values, the ability to distinguish the finest has become detached from the ability to demand it. Amassing physical, financial or even human capital doesn't bring an automatic build-up of cultural capital, and the new rich have constantly struggled to acquire the taste of past, status-based élites. What people spend most on today is by no means what they'll value when the dust settles on the archive, as shown by the number of now-famous artists who died in poverty. A million satisfied buyers can, in the cultural market-place, be totally wrong. So the discerning art lover or music buff must find other ways of transcending majority taste.

Appreciation of great art, and absorption of high-powered knowledge, require separate investment in cultural capital. Accumulation of such capital makes it possible to consume a greater value of cultural products in the time available, just as addition of physical capital makes it possible to produce a greater value of physical products. As with every other form of capital, the cultural type isn't equally spread, but tends to be captured by a privileged élite. But this needn't be, and increasingly isn't, the same élite that concentrates the physical and financial capital. Tension between aristocrats with highbrow tastes and low bank balances, and traders who have acquired worldly wealth but not aesthetic wisdom, is a recurring theme of industrial-age novel plots and plays, of the sort that the aristocrats would attend if only they could still afford it, and from which the traders stay away until Lloyd Webber does a musical version.

Old élites which have lost their economic power don't immediately relinquish their grip on cultural trends. Nor do their newly-rich replacements force them to do so. Their aim is usually to take over that grip, and not just break it. Keen to escape the horse-trading, hand-dirtying world in which their fortune was assembled, they generally aspire to the manners, manors and other ways of life enjoyed by the previous élite. But material wealth does not translate easily to the aesthetic variety. From Dickens's fictional industrialists, unable to square philosophical values with production-line 'facts', to Jeffrey Archer's industrialized fiction, springing a self-made storyteller from the world of noble lords to that of prison dramatists, economic take-off multiplies the incidence of stumbling social climbers with more money than cultural sense.

Capitalism is no more able to deliver full enjoyment on the cultural side than full employment on the economic side. It reproduces known designs until they lose all attraction, but lacks incentives to replace them with the genuinely new. The novelist who starves in a garret, or the artist who can't afford a new brush, is a drain on the economy as well as a burden for partners and relatives. If artists are to survive at all, they must drop the epic ambitions and produce what people want. Many of today's successful novelists – from Salman Rushdie to Fay Weldon – honed their skills and bought their word-processors by writing advertising copy. Too much market pressure – fuelled in part by the effectiveness of that copy – could mean that tomorrow's artists never make that break. Other artforms face a similar danger. As wealthier nations commodify, mass-produce and export artistic and 'knowledge products', these can attain all the mind-shaping and shadow-casting power of military-industrial products. Weapons of mass distraction carry over to post-industrial TV screens what weapons of mass destruction brought to Cold War radar screens.

## Technology to the Rescue

Whereas market-driven globalization steps uneasily into the artist's studio, it has found a welcome reception in the scientist's lab. Confronted by the prospect of getting private funding for their enterprise, artists see the epic poem shelved in favour of advertising copy, the string quartet set to a disco beat, and a corporate logo in the corner of every canvas. Scientists generally see less to lose in corporate sponsorship, and more to gain as their enterprise develops. Traditional university or state support could not afford the vast apparatus required to splice their genes or smash their atoms. Industrial support also opens the prospect of getting intellectual property rights on their results, so adding the fortune of follow-up exploitation to the fame of initial discovery. For every Tim Berners-Lee, willing to forgo patents on the World Wide Web so as to let everyone log on, there may be ten J. Craig Venters, racing to register a patent on the human genome so as to reap a monopoly profit on the use to which it is put.

Industry is keen to fund science researchers, however unfathomable their jargon and eccentric their behaviour, because scientific knowledge now looms large in new product development. Widget designs and workplace layouts can still be reached by real-life trial and error, but it's quicker and cheaper to simulate them on computer. Drugs move faster to market when their effect is tested on screen before being fed to the luckless laboratory rat. Equally importantly, the process of producing that knowledge can be subjected to new industrial management techniques, just as surely as the goods and services which that knowledge will eventually produce. Like commercial production, and unlike artistic endeavour, science is one part inspiration to nine parts refinement and routine slog, with the refiners and sloggers generally getting the money while the inspirers generally wilt under the critical acclaim.

Science's gift to global commerce comes in the form of technological innovation. New products satisfy or bring to light unmet consumer needs, so giving people the incentive to keep buying. New processes cheapen existing products, so help to widen their market. Electronic technologies have been especially productive in both directions, and drove much of the industrial world's most recent burst of growth. Their communication applications have also had important impacts on the way in which other products are traded (insurance down the phone line, electricity over the Internet), and opening up trade in new products (from spare hard disk capacity to organs for spare part surgery) that couldn't be commercialized before.

Microprocessing and electronic networking have made it safer for richer economies to open their borders, by boosting their labour productivity so that higher wages won't mean an immediate outflow of jobs. With enough robotic help, employees exposed to emerging-market competition can keep their generous wage packet, while efficiency gains and savings in non-wage expense keep total (unit) cost down close to low-wage economy levels. Automation may have forced many early retirements from the old assembly-line crowd, but it helped the survivors to increase their output enough to stop cross-border competition from cutting their pay-packet down to size.

As well as being able to afford newer and better machines, the wealthier world can spend more on training its workers to use them, and draw on more advanced transport, distribution and administrative infrastructures with which to carry the cost advantage from factory gate to supermarket shelf. As observed in chapter three, low-wage nations may be able to acquire capital's financial, human and physical forms easily enough, but they need much longer to get the complementary institutional and social forms.

By opening up a new range of service industries, mostly geared to high-income households and up-market producers, computer and communication technology (CCT) has even managed to reverse the financial and physical capital flow, with funds flowing back from emerging economies into those at the technical cutting edge, especially the US. Because first adopters tend to need fistfuls of cash, new CCT-based products are best launched into the countries with the deepest-pocketed shoppers. Because CCTs and the consumer products they power tend to get more valuable the more people want them, they are also best launched where the numbers of such loaded buyers are largest. So it is in the giant shopping mall called the United States that investors see new techs reaping their highest rewards – with the European Union closing the gap as Germans shed their belt-tightening ethic and Britons master the art of living above their means.

CCT products have the added advantage that, when it comes to consumption, the incentive to buy them (and what people pay for them) tends to rise the more widely they are held. Items with rarity value (designer clothing, celebrity autographs), and those subject to physical congestion (cars, ski slopes), see their price premium shrink as they become more commonplace. The more people have them, the less useful they are to the one who buys next. With network goods – phones, games consoles, computer operating systems, join-the-club fashion clothes – usefulness rises with the number of users. So while the first few thousand might have to be given away (to a trend-setting pop idol if not to her fans), the next can be sold at a widening premium as more try to clamber on the bandwagon.

But digital technology, and especially CCT, have done more than just buoy the economy with new lines to spend on. They have changed the way people think, act and work – the cultural context of consumption. This is both a great enablement and a significant constraint for the current globalization phase. Those who have so far used CCT to coordinate protests against the process would do well to consider its potentially more positive effects.

## The Liberating Net

Until a century ago, technology was localizing in its impact. Its transport options knit cities together, but couldn't easily conquer the spaces between them. Its input to weaponry made borders easier to define and defend. The science underlying it assumed that objects and events needed to be close together for cause and effect to run between them. So actions could have only proportionate reactions, their impact remaining local in time and space. There was no instant resonance of one part of the universe with a distant other, no way for time to warp or space to be travelled much faster than light, so no way to cross from one country, continent or constellation to another except with so large a stock of holiday reading that the spaceship couldn't fly.

Then came the moment at which gadgetry went global, tying the world up in a knot of crossing wires and linking waves. Communication technology leapt from telegraph and telephone that could go wherever the cables stretched to voicemail and the Internet, which overcame time-zones just as well as they dispensed with space. Space technology shrank the world by showing that local and global were a bogus distinction, when Earth's fragile blob was viewed against the immensity of interstellar space. And biotechnology showed us how human form and physicality were essentially the same the world over, local difference arising from arbitrary social convention which the universal sciences could usefully overthrow.

Nothing brings home the Internet's impact more effectively than the end of the necessity to leave home. A well-connected household can plug into all of the planet's sales points at the push of a button. Its wilder sights and sounds can be encountered from the safety of a screen. Or rather, all this will be possible, once online retailers get their 'fulfilment' right, and enough cheap broadband arrives to send the pictures at the same speed as the voice.

The 'Net' began as a means to pass around scientific knowledge, and its developers have done their best to keep the sharing spirit alive. Open-system software, and programs whose authors encouraged users to take copies, sped the spread of browsers, hyperlinks and search engines that let cybernauts keep orientated in their deepening information space. Online pioneers measured their profit as added value multiplied by the number of users, transcending industrial pioneers' fixation with those users' ability or willingness to pay.

The Net's commercial users do not deny that market forces could never have produced it. Profit-driven predecessors – electronic data interchange, Compuserve – never made it beyond a small circle with deep enough pockets not to worry about the subscription and narrow enough interests not to mind that so few others were online. But while interactive media are the product of an openness that puts value above market price, the message they carry goes even more in this anti-commercial direction. In reconfiguring the way in which communities connect and talk to each other, new media players have opened up the Net to a whole new ball game.

Telecommunication breaks the arbitrary link between identity and locality. It lets people choose their friends and foes according to common knowledge and interests, not by arbitrary assignment to a common city or street. Sociality becomes a logical function of shared views, not a random result of shared space. Local communities can be enriching, forging links between neighbours who might otherwise never talk. Local spirit can liberate, precisely because it grows from people thrown together here and now, products of an unchosen past and an unconsensual future. But local associations can also oppress those within them who don't reach the group's standards or conform to its norms, and divisively discriminate against outsiders. The tighter-knit they get, the more constricting the social binds become. For every favourite son, the village has a dozen oppressed daughters; for every pristine pillar of the community, a whole row on which the dogs have raised their legs, or disaffected youth discharged their spray-cans.

Cyber-conversation also extracts the substance of identity from the labels that get attached to it. In the newsgroup it doesn't matter who you're talking to, as long as they know what they're talking about. In the electronic market-place you don't mind who you trade with, as long as the buyers have the money and the sellers come up with the goods. In the chatroom you don't know if the pseudonymous suitor is male or female, young or old, tall or short, so all assignatory preconceptions go out of the Windows (95 or above). In cyberspace, no one can hear you scream your unfounded prejudice. But they can recognize your wisdom, whether you're an expert in their area or a casual visitor who just happened to notice where the next big breakthrough lay.

So even as it knits into a global network, communication technology reinforces the conditions for localized interaction, based not on arbitrary location but carefully chosen conversation. Links between people arising from shared belief and interest are much stronger and longer-lasting than those forced by circumstance, whether happy or adverse. Compare the chats you have with someone you meet in a club to those with someone stuck beside you in the lift.

## A Brief Lapse into Political Philosophy

The Internet has become our closest approach to the Veil of Ignorance long dreamed of by philosophers, but never before given operational form. Denied knowledge of their place in time and space by this Veil, people acquire the fair-mindedness and free thinking which social position and historical heritage always previously stopped them from adopting. Not knowing whether they will be rich or poor when the Veil is raised, they opt for an equal distribution of wealth and income. Not knowing which land they'll inhabit, they seek an egalitarian resource spread on an international scale. Not sure whether they will be the ones to sow or to reap, they take the same fairness into intergenerational transfers. So the young build up economic resources generously enough to ensure a viable pension for their parents. They burn up natural resources cautiously enough to ensure that their children still have air to breathe and gas to put in their tanks. What they run up in debts whose costs the next generation must pay, they match with investments whose benefits the next generation can enjoy.

Virtual communities' celebration of anonymity, in contrast to real communities' praise of personalization, is the digital-age version of an ancient conflict between two fundamentally different ways of life. It was previously seen in the eternal war of village against city, bank branch against call centre, expert panel against opinion poll, or happy-family employers against 'flexible' workplaces whose inmates are just passing through. These betray two basically different ways of running a collective effort. In 'familiarized' societies, people place great weight on knowing the identities and abilities of those they deal with, and what someone does is bound up with who they are. In 'functionalized' societies, person and role are kept separate; and because the interaction is one of roles, there's greater efficiency and fairness when we don't know who, or how able, are the other people in the deal.

Functionalized societies can be depicted as de-personalized structures, just like a corporate chart that outlines who does what before anyone's appointed to the actual jobs. The typical functional society was set up as a result of deliberate national identity-building, and has a written constitution. The US is the most prominent example, and also shows most of the other characteristics of societies ordered by function. Because the constitutional plan can be blueprinted, there is potential to reproduce it elsewhere – hence the American proclivity to export their political and lifestyle behaviour (examined further in chapter seven). Because transactions between people are largely anonymous, and cannot be based on knowledge or goodwill of other parties, trustworthiness tends to be gauged by wealth, creditworthiness, or other obvious monetary measures. So functional societies lend themselves to open-market forms of exchange, a significant advantage given the way the world economy has turned.

As a further economic advantage, functional societies work well with detailed division of labour. If technical efficiency requires someone to be a green ink retractable ballpoint pen assembler or cardboard box third flap gluer, they'll do it, if the pay is right, even if their preferred career choice was moral philosophy. And if they quit, because the pay wasn't so good, or gluing down the flap clashed with their libertarian instincts, another less choosy specialist can quickly parachute into their place. When they get away from work, functional societies are also uniquely at home with telecommunication, being happiest with conversation which doesn't need the presence of other parties. And because they've no qualms at talking to strangers, functional societies lead the world in the number and nature of their chat-up lines.

Familial societies, which tend to have evolved their nationhood and based it on impressionistic legal and ethnic distinctions, have great difficulty adapting to a world now being re-fashioned by the functional. The paradigmatic cases are probably China and Japan, where status ascribed through a person's background, known character and career path tends to dominate status achieved by 'objective' measures of competence, or money-making ability. A familial society has difficulty exporting its ways of living or working, because these are so closely shaped around particular people and places, whose like cannot be easily rediscovered abroad. It struggles to establish a detailed division of labour, because people judge others (and themselves) on a skill set too broad to fit easily into one cramped task on a narrow production line. It embraces telecommunication only when designed as an extension of face-to-face conversation. Hence Japan's pioneering role in permanent e-mail and Internet connections, by-passing the dial-up variety on which US online communities grew.

## The Dilemma of Diversity

Fans of the growing global market-place don't have to persuade us that it's the best of all possible worlds. They now have half a million years of human history to back them up. What's around us, they say, is no accident. It's the result of relentless competition that slays any ways of thinking or doing that aren't completely optimal. Today's institutions and ideas survive and thrive because they out-performed the alternatives. To be still around, they have to be the best around. Any superior approach would, by now, have stepped in and stolen the show. So the clamour for change is actually a call to turn the clock back to solutions that were tried and found wanting several centuries ago.

Globalization is a product of this Darwinian, 'retained-because-judged-relevant' logic. Nationally divided systems of culture and commerce were tried, but proved inferior to the border-hopping version. So people who exchanged items and ideas across frontiers came to command more resources, materially and mentally. Their success persuaded others to break down boundaries and adopt a similarly international outlook. Where the globalizers' superior performance didn't push neighbours into opening their borders, superior firepower usually did the trick.

Likewise, the main forces and philosophies driving the present globalization draw praise from their supporters for passing out top of the natural selection test. The shareholder-driven multinational company achieves greater efficiency and faster adaptation than its state- stakeholder- or socially-owned competitors, so outpaces them in the market-place, until they either copy its profit-chasing ways or disappear. The low-taxing, light-regulating, labour-disciplining government gives global business its most fertile breeding ground, so out-prospers more interventionist states, until these copy its techniques or capitulate to its troops. 'Western' machinery, medicine, movies, make-up and maths prove more attractive and effective than counterparts from elsewhere, so find their way onto the library shelves and shop shelves of the world. When free to choose, people grab their global carrier-bag and march into the global market-place. What they come away with is our closest approach to knowing what the world really wants.

Long-distance lines carry messages that break the tyranny of geography. But the lines themselves are also a holdout against the vagaries of history. Network technologies, bringing cross-border links through common technical standards, appear to transcend the 'path dependency' which previously made much technology as country-specific as culture. Telecommunication typifies the new, universalizing technology which works equally well for everyone, and better than what went before. Arrival times may have varied, but every railway carriage now echoes with the chirp of oblivious mobile phone users. After the screech of a car tyre and the creak of a long-shut door, the sound that most aptly conjoins anticipation and trepidation is the crackle of a dial-up connection breaking through into the Net.

Cars, computers, televisions, fridges, microwaves, supermarkets and service stations have similarly seeped into every corner of every culture. Consumer electronics' multilingual instruction leaflets confirm that the globetrotting mission is as much a feature as the digital timer or the self-cleaning screen. ('Do not turn upside down' is printed on the underside of most ready-meals, proving practical jokes to be another technology that crosses cultural divides.) Because these objects are accepted the world over, the infrastructures they spawn also take on universal appearance. Japanese pylons may scrape the sky with a makeshift pagoda, and Ireland may paint its phone boxes green. But more substantial variations now get impaled by the logic that once the right way is found, there's no point casting around for any other ways to do it.

Set-piece strategies work well unless the goal-posts suddenly shift. They're dangerous when there are people on the pitch whose chosen tactic is shifting the goal-posts. Pressure for continuous peak performance exposes markets to the menace of monoculture, in which everyone gets stuck into a way which later ceases to be best. Farmers experience the trap when a tiny bug bites, and each identical field succumbs to a single crop disease. Now computer-makers feel it when suddenly everyone switches from traditional to flat screens, and airlines when one wing fault forces grounding of a whole fleet. The competitive economy's dilemma is that, for efficiency now, it needs to narrow down to one best way of doing things. But for efficiency and growth over time, it needs to keep on finding better ways. The better ways must be chosen from alternatives to the current market leader. But the leader's success lies in snuffing those alternatives out.

## Losing the Argument

Confining the economy to one nation, trade- and capital-protected from the others, gave markets a natural shield from the monocultural menace. When changed conditions made the current way of doing things inefficient, they could look abroad for an effective replacement. The US escaped its two-tonne motor culture with small cars imported from Japan, and is likely to flee from ties to a deadly railway track with high-speed trains from Europe. When one defect ravaged the national monoculture, an insulated store could replenish the stock. France made full use of it when Australian grapes, whose export it had bitterly opposed, were brought back in to replace the vines that a local virus had laid low.

Economies need a source of variety, kept safe from the competitive drive to standardize and generalize. Of the two main ways to do this, neither comes naturally to a world of market-based competition. One solution is to keep the system moderately inefficient, so that alternative ways can survive even though they're not the best right now. The other is to broaden the criterion for efficiency, so that different approaches can be recognized as equally good.

Competitive markets are designed to stamp out inefficiency

by re-allocating resources to their 'best' use, with efficiency strictly defined as bringing maximum profit. So globalization based on them puts the economic system at risk by breaking down the barriers that protected plural thinking. It not only puts the world onto an economic path heading for the economic and environmental buffers of chapter four, but also bricks up the entrance to alternative paths. World politics are in peril from a similar stamping out of national variants. One reason why Europe has retained the vestige of political debate has been its preservation of nation-based parties with their own historical traditions and adjustment to specific local conditions. The longer-unified US showed little remnant of political debate, even before embarking on its latest mass rally for the one best way.

A best-practice spirit led physicists at the end of the 19th century to claim that their own investments had paid off, because they now knew almost all there was to know about their subject. The book of the material universe was virtually closed, and those who came next would have a steady job just teaching it. Luckily, a few of their students went along for the easy ride. By the 1930s they'd discovered that Newton's Laws – seemingly a basis for explaining matter's positioning and movement throughout the universe – broke down at the sub-atomic and super-nova level, and ignored a lot of unpredictable chaos-like conditions in between. The revised book is still being written, and carries over little from the old.

Economists had made a similar claim around the middle of the 20th century, which should have given immediate grounds for suspicion. Even as 'general equilibrium' was being offered as the total picture, deeper probing was unearthing complications that stopped there being a unique one – and a tendency for deviations to compound rather than correct themselves, making equilibrium a purely local and temporary condition. By the 1980s, even economists in the mainstream tradition acknowledged that most of their past 50 years' forecasting efforts had been based on models that invited false correlations between the numbers, and ignored structural change that meant the next event could not be predicted from the last.

Whereas the purchase of new material instruments of production inexorably expands the physical capital stock, investment in new informational instruments of production generally erodes the knowledge capital stock. It had been a token of faith from Europe's 18th-century 'Enlightenment' period onwards that the more we reasoned, experimented and discovered, the more we would know. Exploration and experimentation were viewed as creating an information flow which continuously raised the knowledge stock. Other disciplines, and indisciplines, have similarly learnt that the more they discover, the less they actually know. When man still wielded clubs and lived in caves (with woman waiting patiently in the apartment) most new information added to the stock of human knowledge. Now, with minds full and libraries bulging, and more research time spent debunking old ideas than establishing new ones, it tends to subtract at least as much as it adds. Without this natural safety valve, the world – which funds more science and social research than ever before – would by now know all there is to know. As it is, there is an information overload, but still a very substantial knowledge gap.

## Ignorance is Strength

The more we unearth, the less we know. As a non-material product, knowledge provides a form of growth that counters the 'green' charge of relentless resource depletion. As a product whose stock need not be raised, and can even be reduced, by further investment in its production, knowledge answers the 'red' charge of inexorable demand deficiency. A knowledge economy can potentially beat the under-consumption and overproduction breakdowns outlined in chapter four. Knowledge economies lay the foundation for a style of growth which can keep producing without getting buried under unsold stocks, or poisoned by overflowing dumps.

Globalization has so far been an inadequate escape from these twin breakdowns, because it merely lengthens the journey that's possible before choosing between the evils of over-production or under-consumption. But its economic and cultural effects can provide a basis from which to readjust our expectations, and widen the gap between the collision points.

*Information* is the capitalist system's ideal product, because it represents self-liquidating investment. Its value decays over time, and that decay speeds up in proportion to the rate at which new information is produced. But because so much is needed and so little gets conserved as knowledge capital, information must be continuously produced. This condemns the information-based economy to a resource-expanding growth path, of the type whose bicycle-like instability – needing to keep moving to avoid falling over – was assessed in chapter four. As a counterweight, *knowledge* filters out the information, growing only when there's something genuinely worth adding. So a knowledge-based economy can jump off the run-to-stand-still treadmill, switching to a pedestrian form of growth which allows its passengers to stop and admire the view.

Information is right or wrong. To cope with the overload, what's wrong must at once be dumped. By the time someone sees another interpretation which could mean that it was right, the torn-up printout is buried too deep to be retrieved. In contrast, knowledge has greater or lesser plausibility, but there's little that's absolutely true or false. What's in doubt can still stay in the archive, and there's room there because new research keeps confining old ideas from the main shelves to the basement store. Those stored notes can always be retrieved. This way, a knowledge-based economy, moving beyond the logic of 'economies of scale', can potentially maintain the variety of technique and opinion needed to stave off market monoculture, and keep its politics plural.

Corporate-driven globalization does threaten the production and preservation of knowledge, by treating research as a private investment project and dissemination as a struggle for monopoly market share. But it can also bring escape from these traps by speeding up the fragmentation of capital, from its physical and financial into the new human, social and cultural forms reviewed in chapter three. When capital moves up from commodity to conceptual form, and so returns to being invested in people rather than things, old economy–society relations can be reversed – with commerce drawing on the lessons of culture instead of just drawing in its creations. Anti-globalizers have already played a vital part in this process, by keeping alive economic opinions that ivory-tower consensus would otherwise have driven out of fashion.

Pro-business environmentalists now claim that new technology can multiply our output while reducing our resource use. This would steer us from over-production into under-consumption, unless ways are found to absorb all of the extra output that results. One way, to be tackled through revived global governance institutions, is to transfer purchasing power to poorer countries so that they can help consume the rich world's surplus. But to make this palatable to the privileged, wealthier countries must find a way to achieve additional income growth without requiring more material products. Squeezing more knowledge into those products, and into the process of their production, is potentially a way to do so. The start-of-century slide back into war – the more established way to increase output – by the US, the world's biggest knowledge economy, gives a graphic demonstration of how difficult that potential is to grasp.

## CHAPTER SEVEN

# Them and US

*The high profile of US retailers, banks and restaurant chains abroad makes it easy to view globalization as thinly-disguised Americanization. But Americans are an insular people with no inherent wish to foist their lack of culture on others; and their economy has so hollowed out that the country buys far more goods from the rest of the world than it's able to sell. That voracious material appetite must be financed by capital import, and that's the real reason why the American way is so widely resented. Other nations must assist the US in maintaining a level of life that they can't aspire to, because their own more modest lifestyle would collapse if the biggest spender were ever to tighten its belt.*

Discovering America took several attempts. The Vikings probably went there, but were too embarrassed to admit it. Greeks and Romans reconnoitred but, then as now, found it too much of a cultural and spiritual desert to linger long in. Recent historical evidence suggests wresting the eponym from the glamorous Amerigo Vespucci and giving it to a commercial sponsor of British voyages of discovery, Richard Amerike. For prospecting merchants, the fertile prairies below the equator seemed much more promising than the windswept plains above. Eventually Columbus got quizzed about his late arrival in the Indies, and the secret was finally out. Though even then, having crash-landed in the south, the Conquistadors took a while to work out just what lay above them. Few expected the direction of invasion to be turned round quite so soon.

Defining America similarly defies easy answers. As a promised land, it soon hosted a bewildering cluster of religious sects. As a place to stay, its insistence on four-wheeled transport – to the extent of not even building suburban sidewalks – was soon confusing those visitors whose own language still had the concept of 'stroll'. Some came to suspect that there was nothing to comprehend. The sheer number of its inhabitants seeking the truth – about themselves through analysis, about the world through worship, about others through $50-an-hour private eyes – suggests either unusual levels of demand, or an extremely short supply.

America's recent origin and hazy destination have allowed generations of novelists, Senate sub-committees and psychoanalysts to attempt a diagnosis of American identity, without any risk of finding clear-cut answers. This, after all, would forgo the profit to be made from keeping the controversy going. All sides can at least agree that such a missed money-making opportunity would be wholly out of line with the American Way. But when that Way finally looked beyond its borders, the rest of the world proved to be equally out of step with it.

## Public Enemy: Number One

To outsiders, America's role is much easier to see, by virtue of being impossible to escape. The USA entered the 21st century as the dominant force in global economic, military, cultural and environmental affairs. It has the market that any self-respecting enterprise must sell to, and harbours technologies that most of those enterprises need to buy in return. This gives it the buying power to stamp its own prices on what people offer it, and the borrowing power to procure even the things it can't strictly afford. As ebbing financial sovereignty forces more countries to adopt the dollar, or tie their currencies to it, America's Federal

Reserve can directly dictate the cost of borrowing and pace of development for an increasing proportion of the westward-leaning globe.

US economic ascendancy is made worse by the fact that, although they top the global productivity tables, Americans never seem to work very hard for their world-beating living. Their image abroad is mainly conveyed by films of American suburbia where, between the romantic liaisons, health and beauty fixations, psychotherapy sessions and massive meal-breaks, there scarcely seems time to go to the office or stand at the production line. Whereas Asians carved out their comforts by sheer hard work, and Europeans by their Nobel Prize-winning ingenuity, Americans' luxury lifestyle just seems to be routinely thrust upon them, or at least left on their doorstep by a DHL van.

Resentment at this situation was well illustrated during the great American non-depression of 2001–02. America had been riding a ten-year wave of economic prosperity, bringing jobs to virtually everyone and pushing an increasing number into the millionaire bracket. At the end of such a long and precipitous boom, friends and foes alike expected a crash of equal depth and durability. For a few anxious months, after the close of the Millennium celebrations, it looked as if they'd get it. US shopping malls were empty, its factories closing and offices shedding staff, as capitalism's under-consumption tendencies seemed at last to have found a way into its heartland. But when they looked back over the bleak statistics, Americans discovered that their economy in its 'recession' year had still grown by more than 1%, their disposable incomes by more than 3%, and that the peak of their unemployment was below what Europe suffers in the very best of times. Since it offered the best defence against downturn elsewhere, the rest of the world should have relished the American economy's resilience. They grudgingly did, but not without a twinge of envy of this life of leisure at the centre of the global division of labour.

For those not kept in line by Stateside spending power, its firepower does the job. The US pours more into weapons of war than the ten next biggest defence spenders put together. This helps to keep partners, puppets and pupils of the Pentagon in command of large parts of the world; and ensures that none is out of reach of the lone superpower's nuclear missiles and military propaganda flow. So where the Federal Reserve's writ doesn't run, its military can soon march in to call less financially sensitive satellites to heel.

US monetary and military might help to clear the airwaves for a globetrotting flow of TV news and audiovisual entertainment whose programmes are easily filled by those who exercise that might. So even nations which resist American economic and political hegemony find their front rooms invaded by its images, language reshaped by its buzzwords, and ears retuned by its musical traditions. Hollywood displaces the historians, routinely re-scripting world events until US Marines lead the Normandy landings, Kennedy shoots himself, and stone-age kids with Los Angeles accents keep dinosaurs as pets. Levi's re-educates the world on how to dress, Wal-Mart shows it how to shop. Nothing in the modern world can wholly escape America's influence, except for Henry Kissinger's accent and the rockets that their Missile Defence shield keeps trying to shoot down.

## America the Dutiful

Big nations don't need to be powerful. China has played the role of Sleeping Giant for most of its recorded history, or at least its history as recorded by the people it might tread on if awakened. The nation most worried about China finally throwing its weight around is probably Japan, but its fear is more of being usurped than overpowered. Japan's ability to combine huge economic power with negligible political authority in the world has been mystifying the West's diplomats, and infuriating its military planners, for most of the past (not-so-Pacific) century. Newly industrializing nations can afford to keep their strength to themselves because older powers, notably Britain, insist on punching above their own weight.

As one of the numerically biggest nations – even before it started cloning itself – the US was always going to be among its most powerful. That didn't mean that it had to be so widely disliked. Unlike Britain, its predecessor superpower, America has never clung directly to an empire on which the sun can never set. Unlike Russia, its 20th-century sparring partner, it has never mass-transported its ethnic populations. Unlike China, its most likely challenger for 21st-century supremacy, it has never expelled or imprisoned an entire political opposition. Unlike Japan, which comes closest to its economic wealth, it has never taken microtechnology's miniaturization mission so seriously that callers can't fit their fingers to the keypads of the latest mobile phones.

America's ascendancy upsets those outside because it has achieved all of these effects without indulging in any of the usual despicable causes. It exercises colonial control with only the occasional short-lived invasion, and economic power by imposing day-to-day authority from a respectable distance. Its population moves around, within its borders and across the world, with the application of no force stronger than market force, combined with the unique American appetite for more money, something to video or a glimpse of the British Queen. America can dispense with political opposition because everyone has signed up to the same agenda, seeing no affordable alternative to the system in which everything worth having (including political power) goes to those who pay most for it.

Dislike of the United States is based on a simple contradiction. Its economic strength leaves it uniquely empowered to drive a dismantling of borders which its social tradition makes it wholly unsuited to carry out. A basic inability to globalize crosses paths with a basic indispensability to the globalizing project. When Columbus realized where he'd run aground, he wanted to bring back news of the gold but keep quiet about the people who were wearing it. Americans agreed. The natives were hiding from the rest of the world, the newcomers fleeing from it. Neither wanted to share their new fortune with those that forced them to seek it there. It is the rest of the world that begs Americans to come back and divulge their discoveries, not Americans who beat a path through their forced-open door.

## For Information Only

From the Tokyo shopkeeper confronted with an American Express card to the Tanzanian teacher who gets US junk e-mail, no one is safe from manifestations of the American abroad. Not content to be the place with everything, it is the place that shows its everything to everyone else. But only for inspection. Americans are there, like all experts, to show how it's done, not to give their hosts the ready means to do it. Since a world in which everyone lived America's dream would be unspeakably loud, if not insufferably polluted, it's just as well that these displays are hedged about with stern warnings to viewers and listeners not to try it at home.

The accusation that globalizing business is undemocratic, subverting elected governments and eliminating social choice, is as unintelligible in New York or New Orleans as pavement art is in London, or sliced white bread in Paris. The US view equates financial accounting with political accountability, seeing votes with flourished wallets as complementary to votes by secret ballot. When it crosses the Atlantic or Pacific, just as it earlier diffused across the States, American enterprise makes once exclusive articles available to all. McDonald's democratizes restaurant meals by dropping their prices and simplifying their table manners so that everyone can comfortably eat out. Starbucks makes luxury coffee accessible to all. Charles Schwab brings share-dealing to the masses. General Motors promotes the universal right to drive. With *USA Today*, everyone has an affordable and readable newspaper, to be scanned in the time saved by letting *Reader's Digest* condense the latest book.

Through European or Asian eyes, this is a deceptive generalization. What reaches the masses is not the same as what was earlier enjoyed by the élite. Something is lost in the process of commercial 'democratization', a product's shine rubbing off as it passes into millions of hands. Asian and European minds are attuned to an inverse link between quantity and quality, an abridged-version slide from democratic to demotic. The dissemination of culture or consumption goods beyond the small circle that owns them in the original or can spend an evening discussing them is inevitably equated with dumbing-down. When restaurant arches go gold, what's inside turns to dross.

Proof that Americans share the same doubts is seen from the way in which early-adopters trade up when their first purchase gets too familiar – switching from McDonald's to Wendy's, swapping the Chevy for a Cadillac, transferring to an invitation-only Wall Street firm when Schwab starts to advertize on bus stops. But in principle, if not in practice, the US view is still that what's good for a few gets better as it spreads. 'Network technologies' run in favour of this belief. And having spread the good product news among themselves, it's a natural next step to take the gospel of gadgetry abroad.

Explanations for America's way-of-life evangelism spring easily to mind. Perhaps, as a nation of immigrants, they can't see why their ideas should stop crossing continents just because their originators have briefly settled down. They made the first and fastest move down the Fordist and McDonaldist 'one best way' track, and so must have reached the perfection that other people ought to copy. The US is a hotbed of religions in which those whom the gods save they first make rich, forcing them to impose their role-model in order to stay true to their faith. Perhaps it is a land which, having mass-produced everything it could imagine a society to be built on, sees the mass production of that society as the final inevitable step.

Evangelism appears to have spread from America's religions into its more secular beliefs. But the appearance is deceptive.

The image on offer is more to be admired than acquired. Americans may be convinced that their level and style of living are the best. But that doesn't mean that they wish everyone else to have them. The States' sense of achievement, as well as the world's supply of fresh air and water, would be seriously diminished if everyone could run air-conditioning without an off-switch, or drink five cans of fizzy drink on an average day.

Nor, fortunately, does it mean that everyone else *wishes* to have them. The environmentally-conscious in the rest of the world wouldn't sleep at night – except by dint of carbon monoxide poisoning – if they imagined all other nations burning fuel the way the US does. Denser-populated nations, with no more space in their apartment blocks to store more material objects and no more landfill sites outside in which to dump them, would faint at the thought of US levels of consumption – if they had still had enough open ground on which to fall down.

Proof that the source of national pride need not be put on international parade is starkly provided by Japan – still the Americans' closest economic rival, despite its valiant efforts to fall behind. The Japanese have worked hard to provide for all of their man-made needs themselves, and are often surprised at other nations' indignation at the low take-up of their manufactured exports. Japan sells many of its own goods abroad, but only to buy food, fossil energy, and other commodities that its own land cannot yield. When, as in the past two decades, it regularly exports more than it imports, it invests the resultant surplus abroad so that trading partners can eventually produce at home the products that they currently buy from Japan.

Like Americans, most Japanese believe that they have made the best of what geography and geology gave them. But precisely because what they've built seems such a remarkable achievement, they don't expect it to be reproduced anywhere else. South Korea's efforts to build up Japan-style industrial groups are viewed as amusing, but not to be encouraged. China's construction plans for stock markets, steel mills and car plants are given token amounts of capital investment, but mainly in the way that a householder gives politely to charity when the local collector knocks on the door. Sushi bars and sumo wrestling clubs have, in the past few years, sprung up in the smaller streets of Europe and America's larger cities. But they are mostly the work of local entrepreneurs, with nothing like the worldwide franchising and marketing that spread hamburgers across the Atlantic or football down the Amazon. The world's second richest society lives quietly in the way it feels happiest, without wishing those ways on the rest of the world.

## Defensive Expansionism

Americans would like to be equally insular. Few like to venture abroad, and even fewer like to see their top politicians or big businesses do so. They've even resorted to electing a President whose words need translating into every language, including English, to minimize the chances of dialogue with other world leaders. Far from wishing to rule the world, the US approaches globalization in the same spirit that nuclear weapons were used to promote peace – to make the threat so big, they'll never be forced to carry it out.

Americans are singularly ill-suited to life beyond their borders. Few find their freedom of speech fully expressible in other languages, their fashion sense compatible with other climates, or their health and safety aspirations met in other nations' salons and streets. When President Clinton visited Bangladesh, a nation whose nutrition standards and mortality rates aren't far away from America's, his entourage nevertheless brought its meals, its limousines, its medicine cabinet, and even its bathwater from home.

The ubiquity of American-inspired enterprise abroad is built precisely on this fear of leaving home. If they step outside at all, Americans want to be surrounded by all that is familiar. The plane they fly in and the hotel they stay in must be built and run to American design. There must be food of the type served at the local mall, and drink indistinguishable from what you'd get in a downtown dime store. Any shops they are forced to go into must accept the American dollar, anyone they are forced to speak to must comprehend the American accent. If they must stoop to changing money, they want to recognize the logo on the bank branch. And when they turn on the TV or venture into the concert hall, the only sound and vision they're likely to relish are those they recognize from home.

So while most nations try to send overseas only the things that others are missing, Americans must try to export the things that others already have all around them. Their whole subsistence infrastructure must be reproduced abroad, to avoid a culture shock of insurance-invalidating proportions. This is why the task of converting the cosmos to American burger, soda and sweatshirt brands was taken on with such world-beating alacrity. It was the only way that they could safely board a plane out of the country, when business and pleasure ceased to be containable within their own shores.

Other countries might not have minded this intrusion, if the American way had been convincingly ahead of theirs. But as the oldest and proudest of the contemporary capitalist nations, the US has gone uniquely far in submitting its non-economic activity to commercializing discipline. This is where art has slid furthest from democratic to demotic, and knowledge from in-itself to in-it-for-the-money. In consequence, while other nations selectively export their culture, America indiscriminately exports its lack of culture. One of the culturally least distinguished nations has become the global master at cultural dissemination.

Recipients are, unsurprisingly, perturbed by the arrival of America's cargo culture. It makes little sense for opera-honed musical sense to be battered by Texan hard rock, or gourmet thirst to be slaked on carbonated lime. Parisians watch in dismay at the Disneyfication of their countryside. Istanbullers stare in disbelief at the Starbucks-strangled spanner thrown in their traditional coffee-making works. Their anger spills out in the occasional shattering of shop-fronts and walking-out from the latest Kevin Costner film. But when the protesters return to their muzak-numbed senses, they realize that there is little more substantial they can do to resist. And that the rest of the audience, uncured of their Americanized addictions, never went away.

## The Undeserving Rich

The trouble is, we all need the USA, even if it's a habit that most would like to kick.

Foreign observers used to wonder how America's economy could do so well, when its art and science seemed so relatively under-developed. True, US museums and galleries are well stocked with prize exhibits. But they're usually bought from abroad through superior resources, rather than fashioned on the hard rocks or easels of home. True, the US holds more scientific patents and Nobel Prizes than any other nation. But it does so by buying in the best of the rest of the world's research talent; and by framing intellectual property laws so widely that even natural substances can be patented, and clever turns of phrase given trademark rights.

The US is global Spender of Last Resort. It has to go on buying our goods, even if we have to go on giving it the money to do so. Even if we're not selling directly to Americans, we're probably selling to someone who is. In a world built on market economies with a tendency to over-produce, we have to be thankful for one nation's God-given proclivity to over-consume. As the only nation with a right to happiness built into its constitution, America has chosen to specialize in having a good time. The international division of labour never ruled this out, and a nation tired of working for a living was only too happy to slide horizontally into the role.

Selling to America is the acid test of most manufacturing and service firms' competitiveness. If you can't sell there, there won't be the scale economies to guarantee profitable sales elsewhere. It's a tough job, because Americans train hard for their consumer role, and will reject any product that sticks, squeaks, carries instructions of more than ten syllables, won't run on batteries, and doesn't fit their somehow always slightly larger hands. They also drive a hard bargain, because the US consumes so large a chunk of the world's production that it can generally name any price – except for the subset of goods that it doesn't want to import, because someone at home still makes it and has lobbied Congress to put trade barriers up against it. But even here, raising the tariff isn't always a sign that the US doesn't want the goods. It's usually only exercising another unique big-spender privilege – pushing the world price so far down, because there's too much around when America stops buying, that it can strike an even keener deal when it starts buying again.

Because of its engine-of-demand-growth role, the US can live at other people's expense. Other nations must pay it to consume the goods they can't use up themselves. Once big investors in the rest of the world, Americans have become recipients of other nations' capital on an unprecedented scale. For three decades the US has relied on inward investment flows to finance a deficit on its external accounts, which by 2002 was rising towards 5% of its national output. As a result, the rest of the world had total claims on the US totalling $7.4 billion by the end of 2000, more than 90% of that year's American national income and twice the value of US investment in the rest of the world.

Other countries have had to make this capital transfer because Americans can no longer pay their own way. The huge merchandise export surplus with which America began the post-war period had dwindled by the 1960s, as other nations' rebuilt industrial bases substituted imports and sliced into US export markets. By the early 1970s the US had resorted to printing extra dollars to buy imports that it could not otherwise afford (an inflation-stoking tactic that economists call 'seignorage', and that the rest of us experience when the money we've been paid in is worth progressively less because the price of what we want to buy keeps jumping).

In deference to the many distinct versions into which capital has split (and which chapter three tried to pull together), the US does not stop at importing just the physical and financial varieties. It draws almost as heavily on the world's supply of human and intellectual capital. Without a plentiful supply of foreign physicists and software engineers, America would have struggled to keep its new technology lead. Silicon Valley is reported to have become so multilingual that programming language is the only one that its inhabitants have in common. Wall Street must similarly buy in staff who, when the money talks, can reply in its local language. US universities' science and maths departments have long depended on foreign students to fill their courses, as the locals drop out or (in Harvard, and) drift into the business or law school. The presidency is probably the only skilled American job still always to be filled by a US citizen. But in 2000 even this post passed to a man who gave every appearance of speaking English as a second language.

From the perspective of the nations from which these high-powered migrants come, America's import of capability is a damaging brain drain. While some of the migrants send home big remittances during their time abroad, and eventually return with skills and knowledge that they can re-import, many choose to stay in the States, their skills forever lost to those who nurtured them. Others try to return, but find themselves forced to flee abroad again – pushed out in part by rejection and ridicule of the Americanisms that they bring home.

But Americans view the drain as a result of free choice – an easy justification to rely on, for a nation which usually sets the rules by which such choices are made. When the rules go against it, the US has the privilege of changing them. It effectively junked the post-war fixed exchange rate system by ending convertibility of dollars into gold in 1971. The world had to live with it, because their own economies would have been taken down by any new American plunge into economic isolation. It also had to live with the removal of capital controls that followed the flotation of the dollar, setting off the large cross-country flows of capital that have been occurring since. The US pushed for this precisely because it knew – even as Kissinger was proclaiming the conquest of world hunger by the 1980s – that the biggest recipients of the newly-moving capital would be those who already had the most. By turning the IMF – set up to help nations with liquidity problems stabilize their exchange rates – into a guardian of free capital movement, America found a new route to the purchasing power that it could no longer generate by productive endeavours at home. The US remained an engine of world growth, as recession-fighting necessity forced trade partners to club together to supply the fuel and spare parts themselves.

America's ability to import capital, on a scale denied to other countries even on the OECD rich list, was acquired at least partly on merit. Its combination of industrial restructuring, technological innovation, macro-economic management and aggressive trade policy has persuaded the world that the likely returns on US-targeted investment are higher, and risks lower, than for comparable investment even in regions where capital's scarcity ought to make it more highly prized. But there is also a large unearned element in the US ability to capture others' capital and live on the unearned income that results. America's attractiveness rests in large part on its possessing the world's reserve currency (and so being able to borrow abroad without currency risk); on having the world's largest single consumer and investment goods markets (which inward capital and labour flows now augment with no additional effort); on its geographical equidistance between Europe and Asia (a gift of ancient tectonic drift); on its near-fluency in the English language (a table-turning colonial inheritance); and on its own insatiable appetite for imports (which allow it to set its own prices through monopoly buying, optimal tariffs, and judicious shows of force whenever embargo threatens).

The incoming money hasn't just been handed over as foreign aid. It has been invested or lent, so other nations hold an increasing share of American industrial equity and have increasing claims on its income via corporate and federal government debt. But America is uniquely empowered to shrink its debts, by printing more of the dollars in which they are denominated. This devalues the dollars that they pay back, and inflates world prices so that the real value of the debt continuously declines. Because it has drawn so heavily on the rest of the world, the US enjoys the same privilege as any private client who draws too heavily on shareholders and creditors. The holders of US shares cannot sell them without sinking the stock market, and shrinking their own fortune. The holders of US debt cannot call it in without forcing bankruptcies that will stop their capital from ever coming back.

If the world were to stop giving Americans the money to live above their means, the Untied States would have to ratchet down expenditure until exports were enough to pay for imports, and all investment was paid for by domestic saving. But the resultant turn-around in the US current account would give large parts of the world nowhere to sell their exports. And the necessary rise in US interest rates would undermine the growing number of economies who have adopted or tied their currencies to the dollar, and whose borrowing costs the US can thus dictate. America may have put itself at the centre of a trade and capital flow pattern that inverts globalization's original promise of investing most in those with least. It can do so on the basis that without it, few would have anything. Injustices in the present situation cannot be remedied without a major redirection of world capital, requiring significant demand-boosting efforts to shield the rest of the world from America's 'adjustment'. So far, the world has found it easier to go on buying America's burgers than to share its cold turkey.

## Top Gun

In return for its largesse, the US doesn't just give the world inescapable plastic toys and a lot of retail outlets. America has also become the world's policeman, stepping up and spreading out its defence preparations after the Cold War even as other countries were treating it as a chance to mothball the tanks. For those too impoverished or abstemious to make use of American products, there's always the American war-machine to rely on, if they play the right anti-communist or anti-terrorist cards.

Many nations feel threatened by US military ascendancy, and some have suffered at the warm end of its gun barrels. But this is to overstate America's malevolence. Many others feel shielded by US planes and ships that profess to be on their side. This is equally to overstate its magnanimity. All that America is doing with its huge military machine is defending its own home patch. The trouble is, with such a massive material appetite and such a depleted industrial base, the lines of defence for that home patch now run all over the world.

The oil cannot safely flow from Central Asia and the Gulf without the installation and continuous maintenance of a chain of puppet rulers. The bananas can't arrive from Central America without a five-year trade war with the European Union. The airspace of the Northern hemisphere must be kept open for American private jets, that of the Southern hemisphere for American pollution clouds. Coffee, uranium and training-shoe supplies must be brought from ever further afield, because the existing supplies keep drying up, despite exhaustive efforts to farm, mine and make them sustainably.

## Bottom Dollar

Two centuries on from its fateful rediscovery, America's role as a nation is no clearer. It is still a place where some people do not think twice about taking the elevator to a tenth-floor gym so they can walk on an indoor conveyor belt, or drive across town to drop one glass jar in the bottlebank for the sake of the environment. It still seeks Americanness in added value, through shared possession of shop-bought quantities, when most others try to trace national identity to inherent values, possessed regardless of a person's state of wealth.

America's role in the world is uncomfortably clearer. Today's global economic and military machine is indeed constructed around it, even though half of the multinationals that drive it are now European, and most of the capital that fuels it comes from Asia. But this does make globalization a figleaf for the export of American culture. It has so little of this, or anything else, left to sell. The outward march of General Motors and Colonel Sanders is less an invasion of the world than Americans' attempt to keep it at a safe distance. As long as there are freeways and fried chicken when Stateside visitors require them, the rest of the world needn't dance to Detroit or Kentucky tunes. That first Thanksgiving was for an *end* to the migrants' trouble-fuelled travels, not for a stop-over on some round-the-world mission. The money that other countries channel into American banking, military and retail services is a subscription fee for its continued involvement in the world, not a bribe to make it stay at home.

Americans send their commodities and culture abroad because they prefer to stay at home. Although proponents of their millennial 'anti-terrorism' drive express abhorrence of approaches to life that reject American moral and money-making values, the more typical emotion is bewilderment. If Europeans wish to tax themselves in order to fund collective healthcare, or eat the food that nature intended, or put their fuel prices up so as not to live under a pollution cloud, they are perfectly free to do so. These differences actually help to reinforce American confidence in its technological and ethnological uniqueness. There are frictions when rules and conventions stop US corporations from setting up or selling in the euro-zone. But ceasefire can be swift once America's weapons are seen to be trained on home ground, and its rules designed to stop domestic battlefields from collaterally damaging the foreign training pitch. Other nations can remain free to go their own way, so long as the American Way escapes unharmed.

Therein lies the catch which, by early 2002, had drawn the US into spy-plane confrontation with China, war in Afghanistan, new Iraqi invasion preparations and uneasy spectatorship on a bloodbath in the Middle East. What the rest of the world does differently has a powerful impact on American interests. Globalization has breached an insulating wall between contrasting worlds, forcing once concealed disputes into the open.

Within nations, this confrontation happened with industrialization, when landlords and tenants who had once co-existed in open fields were shoved onto opposite sides as employer and worker in closed factories' disputed terrain. Extreme inequality makes for peace, because the losers are too far below to see the winners or throw stones at them. The explosion comes when the dividing line narrows, to the point where those in the engine-room glimpse the opulence of the state-rooms above and decide to clamber up to them. The same awakening is now playing out between nations, on a world scale, as technology, commerce and electronic media release sovereignties from solitary confinement. The strongest motivation for corrective action is held by the planet's poorest, for whom the disparity between hunger and hangover, AIDS deaths and cosmetic surgery, dirt track and motorway is most stark. Those with most reason to burn with a sense of injustice lack the firepower to resist it, as the Middle East's al-Qaeda network and Mexico's Zapatista rebels are among the most recent to discover. The strongest manifestation of the new dividing line is between Americans who seem to have it all, and other wealthy nations destined to understudy a leader which they can never quite overtake. The rumbling discontent is only made worse by Americans' inability to turn their material wealth into emotional happiness, the poor little rich country which only gets harder to deal with when it sees that the expected sympathy isn't there.

Shared commercial and security interests mean that the dollar-, yen- and euro-zones cannot simply agree to disagree.

They must try to resolve their differences, and evolve the extra global rule-making and refereeing arrangements that an interconnected world system still palpably lacks. The eternal threat of retreat into regionalism – Europe re-expanding to its east, America looking south, Japan turning inwards – shows how much easier it is to bridge large gaps regionally than narrow gaps globally. In the bleakest times, staying alive is all that matters, and agreement is easy on how it should be done. The gulf between rich and poor nations is essentially one of quantity: income, capital, material resources. The divide within the rich world runs deeper, to qualitative issues of how and why to live. Theirs is the mutual distrust that grows from mutual interdependence, the periodic fury that always flares between inseparable companions. For the two centuries since it lost control, Britain has been divided from America by a common language. Now other industrial nations are stumbling on similar shared ground, and wondering whether it will become the scene of barbecues or battles.

A natural instinct is to fight any enemy that gets too close, and this is the urge that anti-globalizers tend to follow. But the world has far more to fear from US isolation than US domination. Its best hopes of survival rest on extending American cultural imperialism from designer jeans and redesigned genes to its wider military–industrial complexities. The more their companies venture abroad and have to adapt to local variation, the more the merits of diversity will be messaged home in the business language that America understands.

Although President Bush announced a 40% steel tariff in early 2002, the sanction was far from across-the-board: exemptions were needed for all of the strands and shapes of the metal that American firms were making or sourcing abroad and sending back to the US. The more America gets mixed up with the rest of the world, the less risk the world runs of getting a free gift from those Pentagon-shaped silos. If that means having our languages strangled by American song lines, as well as our economies tangled in American supply lines, the cushioned widescreen of American engagement is still preferable to the bunkered gunsight of American retreat.

## Parting Shot

Not content with rewriting the world's understanding of economics, America also invented the business school. Graduates of these institutions are now not only climbing the corporate ladder, but jumping across to top jobs in government and its regulatory agencies. While the first President George Bush used the CIA as his launch-pad to the White House, the second began his much more perilous ascent with a Harvard MBA. In America, B-school tutors are an understandable mix of proud ex-practitioners, earnest scholars and cheerleaders of the managerial art. On arrival in Europe, however, the concept has made a shift. While most university economics departments have firmly rejected any lecturers and set texts that step even briefly away from free-market faith, business schools have become a recruiting ground for heretical economic views.

If you want to find dedicated followers of Marx, Mao, Foucault and other enemies of capitalism, look no further than your local Management Studies centre. At some UK business schools, half of the staff are proudly authoring explosive critiques of their company sponsors; the other half giving lectures that leave aspiring executives in no doubt about the ecological evil and social destructiveness of the firms they plan to run.

Money drives this curious cross-over, but not in the predictable way. An orthodox economics that condemns state intervention could only exist with state support, since it is not something that free markets have ever been willing to pay for. Where private funds flow into research, it is of the critical, anti-capitalist variety, dwelling on the deficiencies of the market and the boom–bust tendencies of any system that gives it free rein. Noreena Hertz, Laura Tyson, Lester Thurow and Geoff Hodgson are among the one-world-wary heavyweights who assail free global markets from the safety of a leading business school. Inexplicable to the world of stark anti-global opposition, this trend gains more clarity on closer inspection of the twists in global scepticism's supply chain.

Multinational companies (as we saw in chapter one) must inhabit a planet made economically unstable and environmentally unsustainable by their uncontrolled activity (chapters two and three). Governments' failure to resolve these problems nationally (chapter five), and their resort to actions which recreate them internationally (chapter three), have forced big business to find its own solutions to over-production and under-consumption (chapter four). Corporations have done so by setting up central planning systems that modify, manage or sidestep the market – far more effectively than any form of planning that communists or other anti-capitalists could dare contemplate. If we seek to understand the troubles it faces, and to appreciate the solutions it applies, then it appears wholly logical for business to look to schools whose star performers condemn free markets and despair of *laissez-faire*.

As borders dissolve, a better shape of world can crystallize, if we level the crucible and mix the appropriate solutions. Following big business down its market-ending, nation-transcending path has as great a chance of achieving this as forcing states to resume thinking locally in a world that transacts globally. Globalization is the natural condition of a minor planet, from which politics dislodged it, and to which culture and economy now have the potential to return it. We are less likely to get there by following micro-state-builders back to grass roots than by racing with the multinationals to the ends of the earth.

# Bibliography


Barber, Benjamin. *Jihad vs. McWorld: how globalism and tribalism are reshaping the world*, New York: Ballantine, 1996.

Burtless, Gary and Litan, Robert. *Globaphobia Revisited: open trade and its critics*, Washington DC: Brookings Institution, 2001.

Castells, Manuel. *The Rise of the Network Society*, Oxford: Oxford University Press, 2000.

Desai, Meghnad. *Marx's Revenge*, London: Verso, 2001.

De Soto, Hernando. *The Mystery of Capital*, London: Black Swan, 2001.

Eatwell, John and Taylor, Lance. *Global Finance at Risk: the case for international regulation*, Cambridge: Polity Press, 2000.

Frank, Thomas. *One Market Under God*, London: Secker and Warburg, 2001.

Friedman, Thomas. *The Lexus and the Olive Tree*, New York: HarperCollins, 2000.

Gilpin, Robert. *The Challenge of Global Capitalism: the world economy in the 21st century*, Princeton: Princeton University Press, 2000.

Greider, William. *One World, Ready or Not*, London: Touchstone, 1998.

Hertz, Noreena. *The Silent Takeover*, London: Heinemann, 2001.

Hirst, Paul and Thompson, Graham. *Globalization in Question*, 2nd edn, Cambridge: Polity Press, 1999.

Hodgson, Geoffrey. *How Economics Forgot History*, London: Routledge, 2001.

Keohane, Robert and Nye, Joseph. *Power and Interdependence*, London: Longman, 2000.

Klein, Naomi. *No Logo*, London: Flamingo, 2001.

Micklethwait, Jonathan and Wooldridge, Adrian. *A Future Perfect: the challenge and hidden promise of globalization*, London: Random House, 2001.

Minsky, Hyman. *Stabilizing an Unstable Economy*, New Haven CT: Yale University Press, 1986.

Ritzer, George. *The McDonaldization of Society*, Pine Forge CA: Sage, 2000.

Rodrik, Dani. *The New Global Economy and Developing Countries: making openness work*, Policy Essay 24, Washington DC: Overseas Development Council, 1999.

Rugman, Alan. *The End of Globalization*, London: Random House, 2001.

Sardar, Ziauddin and Wyn Davies, Merryl, *Why Do People Hate America?*, Cambridge: Icon Books, 2002.

Shipman, Alan. *The Market Revolution and its Limits: a price for everything*, London: Routledge, 1999.

Short, John Rennie. *Global Dimensions: space, place and the contemporary world*, London: Reaktion Books, 2001.

Standage, Tom. *The Victorian Internet*, London: Orion, 1999.

Soros, George. *The Crisis of Global Capitalism*, New York: Little, Brown, 1998.

Tanzi, Vito and Schuknecht, Ludger. *Public Spending: a globalist perspective*, Cambridge: Cambridge University Press, 2000.

Thurow, Lester. *Head to Head*, London: Nicholas Brealey, 1993.

Tyson, Laura. *Who's Bashing Whom? Trade conflict in high technology industries*, Washington DC: Institute of International Economics, 1992.

Weiss, Linda. *The Myth of the Powerless State*, Cambridge: Polity Press, 1998.

Whalen, Charles, ed. *Political Economy for the 21st Century*, New York; M.E. Sharpe, 1995.

Yergin, Daniel and Stanislaw, Joseph. *The Commanding Heights: the battle between government and marketplace that's reshaping the modern world*, New York: Simon and Schuster, 1998.

# Index

## Chomsky and Globalisation

*Jeremy Fox*

Noam Chomsky, hailed by some as the 'Einstein of modern linguistics', is equally well known to others as an uncompromising political dissident and social critic.

This book examines Chomsky's libertarian views on global economic hegemony and the new world order. His position is an unusual one: he feels that 'free trade' is not free at all – rich powers ignore its rules in order to subsidise their big companies and only the indebted Third World countries are obliged to obey. Thus, on the ill-balanced scales of global business, the favoured Euroamerican élites must inevitably grow richer, while the rest of the world could revert to the conditions of Blake's 'dark Satanic Mills'.

**Icon Books UK £4.99**
**Canada $9.99**
**Totem Books USA $7.95**

ISBN 1 84046 237 X

## Why Do People Hate America?

*Ziauddin Sardar and Merryl Wyn Davies*

The economic power of US corporations and the virus-like power of American popular culture affect the lives and infect the indigenous cultures of millions around the world. The foreign policy of the US government, backed by its military strength, has unprecedented global influence now that the USA is the world's only superpower – its first 'hyperpower'.

America also exports its value systems, defining what it means to be civilised, rational, developed and democratic – indeed, what it is to be human. Meanwhile, the US itself is impervious to outside influence, and if most Americans think of the rest of the world at all, it is in terms of deeply ingrained cultural stereotypes.

Many people do hate America, in the Middle East and the developing countries as well as in Europe. This book explores the global impact of America's foreign policy and its corporate and cultural power, placing this unprecedented dominance in the context of America's own perception of itself. Its analysis provides an important contribution to a debate which needs to be addressed by people of all nations, cultures, religions and political persuasions.

**Icon Books UK £7.99**

ISBN 1 84046 525 5